Deepwater Horizon Oil Spill
Phase I Early Restoration Plan and
Environmental Assessment

Prepared by the Deepwater Horizon Natural Resource Trustees from

State of Alabama (Department of Conservation and Natural Resources; Geological Survey of Alabama)

State of Florida (Department of Environmental Protection; Fish and Wildlife Conservation Commission)

State of Louisiana (Coastal Protection and Restoration Authority; Department of Environmental Quality; Department of Wildlife and Fisheries; Department of Natural Resources; Oil Spill Coordinator's Office)

State of Mississippi (Department of Environmental Quality)

State of Texas (Texas Commission on Environmental Quality; Texas General Land Office; Texas Parks and Wildlife Department)

Department of the Interior

National Oceanic and Atmospheric Administration

(This page intentionally left blank.)

Executive Summary

Introduction

The Gulf of Mexico is a priceless national treasure. Its natural resources – water, fish, beaches, reefs, marshes, oil and gas – are the economic engine of the region. The Gulf of Mexico is likewise vitally important to the entire nation as a bountiful source of food, energy and recreation. The Gulf Coast's unique culture and natural beauty are world-renowned. There is no place like it anywhere else on Earth.

On April 20, 2010 the eyes of the world focused on an oil platform in the Gulf, approximately 50 miles off the Louisiana coast. The mobile drilling unit *Deepwater Horizon*, which was being used to drill an exploratory well for BP Exploration and Production, Inc. (BP), violently exploded, caught fire and eventually sank, tragically killing 11 workers. But that was only the beginning of the disaster. Oil and other substances from the rig and the well head immediately began flowing unabated approximately one mile below the surface. Initial efforts to cap the well were unsuccessful, and for 87 days oil spewed unabated into the Gulf. Oil eventually covered a vast area of thousands of square miles, and carried by the tides and currents reached the coast, polluting beaches, bays, estuaries and marshes from the Florida panhandle to west of the Mississippi River delta. At the height of the spill, approximately 37% of the open water in the Gulf was closed to fishing. Before the well was finally capped, an estimated 5 million barrels (210 million gallons) escaped from the well over a period of approximately 3 months. In addition, approximately 771,000 gallons of dispersants were applied to the waters of the spill area, both on the surface and at the well head one mile below. It was an environmental disaster of unprecedented proportions. It also was a devastating blow to the resource-dependent economy of the region.

While the extent of natural resources impacted by the *Deepwater Horizon* oil spill and response (collectively, "the Spill") is not yet fully evaluated, impacts were widespread and extensive. The full spectrum of the impacts from this spill, given its magnitude, duration, depth and complexity, will be difficult to determine. The trustees for the Spill, however, are working to assess every aspect of the injury, both to individual resources and lost recreational use of them, as well as the cumulative impacts of the Spill. Affected natural resources include ecologically, recreationally, and commercially important species and their habitats across a wide swath of the coastal areas of Alabama, Florida, Louisiana, Mississippi, and Texas, and a huge area of open water in the Gulf of Mexico. When injuries to migratory species such as birds, whales, tuna and turtles are considered, the impacts of the Spill could be felt across the United States and around the globe.

The Role of the Trustees

Under the Oil Pollution Act (OPA), which became law after the 1989 Exxon Valdez oil spill, the federal government, impacted state governments, federally recognized Indian tribes and foreign governments act as "trustees" on behalf of the general public. Trustees are charged with recovering damages from the parties responsible for oil spills to restore injuries to the public's natural resources. Trustees assess the nature and extent of natural resource injury and develop and implement a plan for the restoration, rehabilitation, replacement, or acquisition of the equivalent of the injured natural resources and services those resources provide under their trusteeship. The *Deepwater Horizon* Trustees (Trustees) are:

- the United States Department of the Interior (DOI), as represented by the National Park Service, United States Fish and Wildlife Service, and Bureau of Land Management;
- the National Oceanic and Atmospheric Administration (NOAA), on behalf of the United States Department of Commerce;
- the State of Louisiana's Coastal Protection and Restoration Authority, Oil Spill Coordinator's Office, Department of Environmental Quality, Department of Wildlife and Fisheries and Department of Natural Resources;
- the State of Mississippi's Department of Environmental Quality;
- the State of Alabama's Department of Conservation and Natural Resources and Geological Survey of Alabama;
- the State of Florida's Department of Environmental Protection and Fish and Wildlife Conservation Commission; and
- for the State of Texas: Texas Parks and Wildlife Department, Texas General Land Office and Texas Commission on Environmental Quality.[1]

The Trustees began working together in the early days of the Spill. The result has been an unprecedented state-federal collaboration, with a unity of vision and purpose, and a strong desire by all the Trustees to act as quickly as possible to restore the Gulf. Trustee efforts to assess the injuries to natural resources began within hours of the explosion and continue to the present.

The Trustees uniformly believe that restoration of the natural resources in the Gulf must begin as soon as possible. This Phase I Early Restoration Plan and Environmental Assessment (ERP/EA) contains the initial plan for the first of a long series of restoration actions that will be undertaken by the Trustees, paid for by those responsible for injuries to natural resources and the services they provide, representing the first step on the road to a full recovery for the region. The ultimate goal of the Trustees is comprehensive and long lasting repairs to the Gulf ecosystem, and the communities that depend on it, to the condition they would have been in if there had never been a spill, as well as to compensate the public for its lost use of the resources during the time they were injured.

From the outset, the Trustees expected that the restoration of resources injured by the Spill would be a massive undertaking, and that during the assessment, injuries would continue to accrue. The Trustees decided that because of the pervasive and ongoing nature of the damages to natural resources in the region, it would be in the best interest of the public to accelerate restoration and begin implementing projects, if possible, even before completion of the full damage assessment. The Trustees approached BP in the fall of 2010, and negotiations on an early restoration fund commenced. Exactly one year after the explosion on the *Deepwater Horizon* rig, the Trustees and BP entered into an unprecedented agreement whereby BP set aside one billion dollars to fund early restoration projects agreed to by BP and the Trustees, incorporating public review.

---

[1] The Department of Defense (DOD) is also a trustee of natural resources associated with DOD-managed land on the Gulf Coast, which is included in the ongoing NRDA, but DOD is not a signatory of the Framework Agreement nor a participant in this Phase 1 Early Restoration Plan.

This early restoration agreement, known as the "Framework Agreement"[2], represents the initial step toward the restoration of natural resources injured by the *Deepwater Horizon* spill. It is a down payment against the ultimate claim for damages from the Spill. The Trustees expect to be able to fund more early restoration projects in addition to this initial set. The Trustees continue to assess the injuries to natural resources and services resulting from the Spill and pursue the ultimate claim for damages. Restoration work will take many years to complete, and long term monitoring and adaptive management of the Gulf ecosystem will likely continue for decades until the Trustees can be certain that the public has been fully compensated for its losses.

Early Restoration Project Selection

Following signature of the Framework Agreement, the Trustees invited the public to provide early restoration project ideas and proposals. The Trustees received hundreds of proposals, which were made publicly available at http://www.gulfspillrestoration.noaa.gov/restoration/give-us-your-ideas/view-submitted-projects/. The Trustees implemented a project selection process to evaluate proposals and ensure that restoration would begin as soon as possible. Figure ES-1 depicts the general selection process, which included project solicitation, project screening and identification, negotiation, public review and comment, and final selection.

The Trustees evaluated potential early restoration projects using criteria included in applicable damage assessment and restoration regulations and programs, the Framework Agreement, and factors that are otherwise key components in planning early restoration. Under OPA regulations, restoration alternatives are evaluated with regard to:

- The cost to carry out the alternative;
- The extent to which each alternative is expected to meet the Trustees' goals and objectives in returning the injured natural resources and services to baseline and/or compensating for interim losses (the ability of the restoration project to provide comparable resources and services, that is, the nexus between the project and the injury);
- The likelihood of success of each alternative;
- The extent to which each alternative will prevent future injury as a result of the incident, and avoid collateral injury as a result of implementing the alternative;
- The extent to which each alternative benefits more than one natural resource and/or service; and
- The effect of each alternative on public health and safety.

Under OPA regulations, if the Trustees conclude that two or more alternatives are equally preferable, the most cost-effective alternative must be chosen.

In addition, the Framework Agreement provides that projects:

- Contribute to making the environment and the public whole by restoring, rehabilitating, replacing, or acquiring the equivalent of natural resources or services injured as a result of the Spill, or compensating for interim losses resulting from the incident;

---

[2] http://www.restorethegulf.gov/sites/default/files/documents/pdf/framework-for-early-restoration-04212011.pdf.

- Address one or more specific injuries to natural resources or services associated with the incident;
- Seek to restore natural resources, habitats, or natural resource services of the same type, quality, and of comparable ecological and/or human-use value to compensate for identified resource and service losses resulting from the incident;
- Are not inconsistent with the anticipated long-term restoration needs and anticipated final restoration plan; and
- Are feasible and cost-effective.

The Trustees also took into account several practical considerations that, while not legally mandated, were useful and permissible to help screen the large number of potential qualifying projects. For example, Trustees:

- took into account how quickly a given project could begin producing environmental benefits;
- sought a diverse set of projects providing benefits to an array of greatly injured resources;
- focused on types of projects with which they have significant experience, allowing them to predict costs and likely success with a relatively high degree of confidence and making it easier to reach agreement with BP on the restoration benefits estimated to be provided by each project (referred to as "Offsets"); and
- gave preference to projects that were closer to being ready to implement.

The Trustees acted promptly to identify project proposals that met the selection criteria, and then narrowed the potential project list down to an initial group to move forward into discussion with BP on cost and Offsets. The Trustees and BP came to preliminary agreement on a set of proposals, which the Trustees proposed as Phase I projects in a Draft Phase I ERP/EA released for public comment in December, 2011.

Selected Projects

Consistent with OPA and the National Environmental Policy Act, the Trustees considered public comment prior to final selection of Phase I projects. A summary of comments on the Draft Phase I ERP/EA, Trustee responses to comments, the final selected list of Phase I projects, as well as environmental assessments of potential impacts from those projects are included in this ERP/EA. In addition, this ERP/EA includes a description and quantification of the Offsets preliminarily agreed to by BP and the Trustees.

This ERP/EA consists of eight projects listed in Table ES-1 and more fully described in this document. They address an array of injuries and are located throughout the Gulf (Figure ES-2). Specifically, this plan includes two oyster projects, two marsh projects, a nearshore artificial reef project, two dune projects, and a boat ramp enhancement project. These projects address injuries in four of the five impacted states, on the coast and offshore, to mammals and marine organisms, and/or compensate for lost recreational opportunities for the public. While this plan includes a suite of projects, each project was viewed and evaluated as independent from the others. This ERP/EA does not attempt to quantify the injury to natural resources; instead it outlines a set of projects which will accelerate meaningful restoration in the Gulf while the full assessment and restoration planning process continues.

Next Steps

This ERP/EA serves as the Trustees' final selection of Phase I early restoration projects, taking into account the suite of potential projects proposed, the NRDA and Framework Agreement process, and public comment on the Draft Phase I ERP/EA. Per the Framework Agreement, the Trustees will move forward with agreements with BP to fund projects and commence implementation, as described in more detail throughout this document. Updates on the progress of project implementation will be available at http://www.gulfspillrestoration.noaa.gov.

Projects selected in this ERP/EA represent only the first phase of the early restoration process. The Trustees continue to evaluate additional projects already submitted by the public for consideration, as well as any new projects as they are received, with the intent of proposing additional projects until funds made available under the Framework Agreement are exhausted. It is important to emphasize that restoration proposals developed pursuant to the Framework Agreement are not intended to provide the full extent of restoration needed to satisfy the Trustees' claims against BP. At the end of the NRDA process, the Trustees will credit all the Offsets identified for approved early restoration projects against their assessment of the **total** injury for the Spill. Restoration beyond early restoration projects will be required to fully compensate the public for natural resource losses from the Spill and will continue until the public is fully compensated for the natural resources and services that were lost as a result of the Spill.

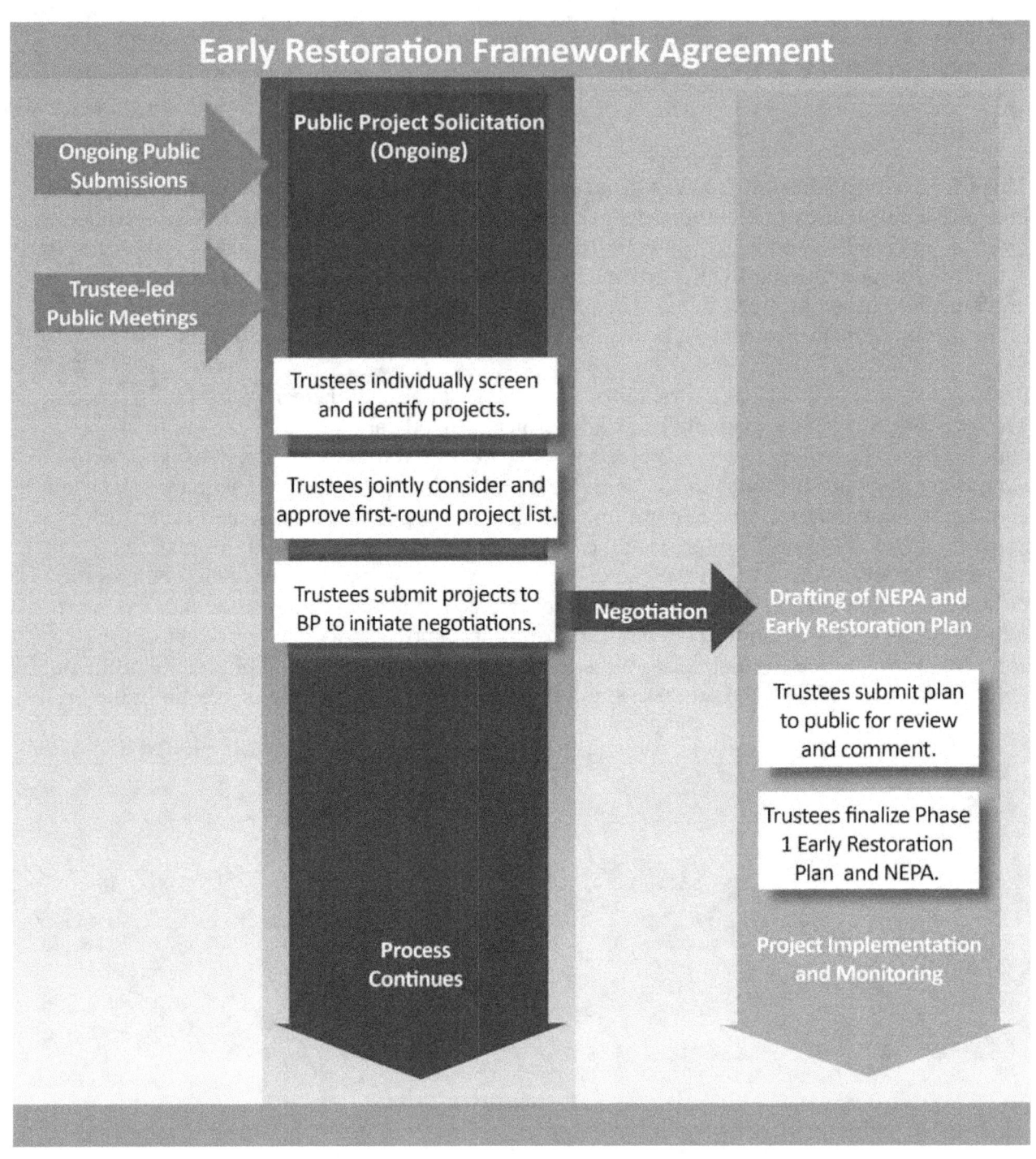

Figure ES-1. General Early Restoration project selection process.

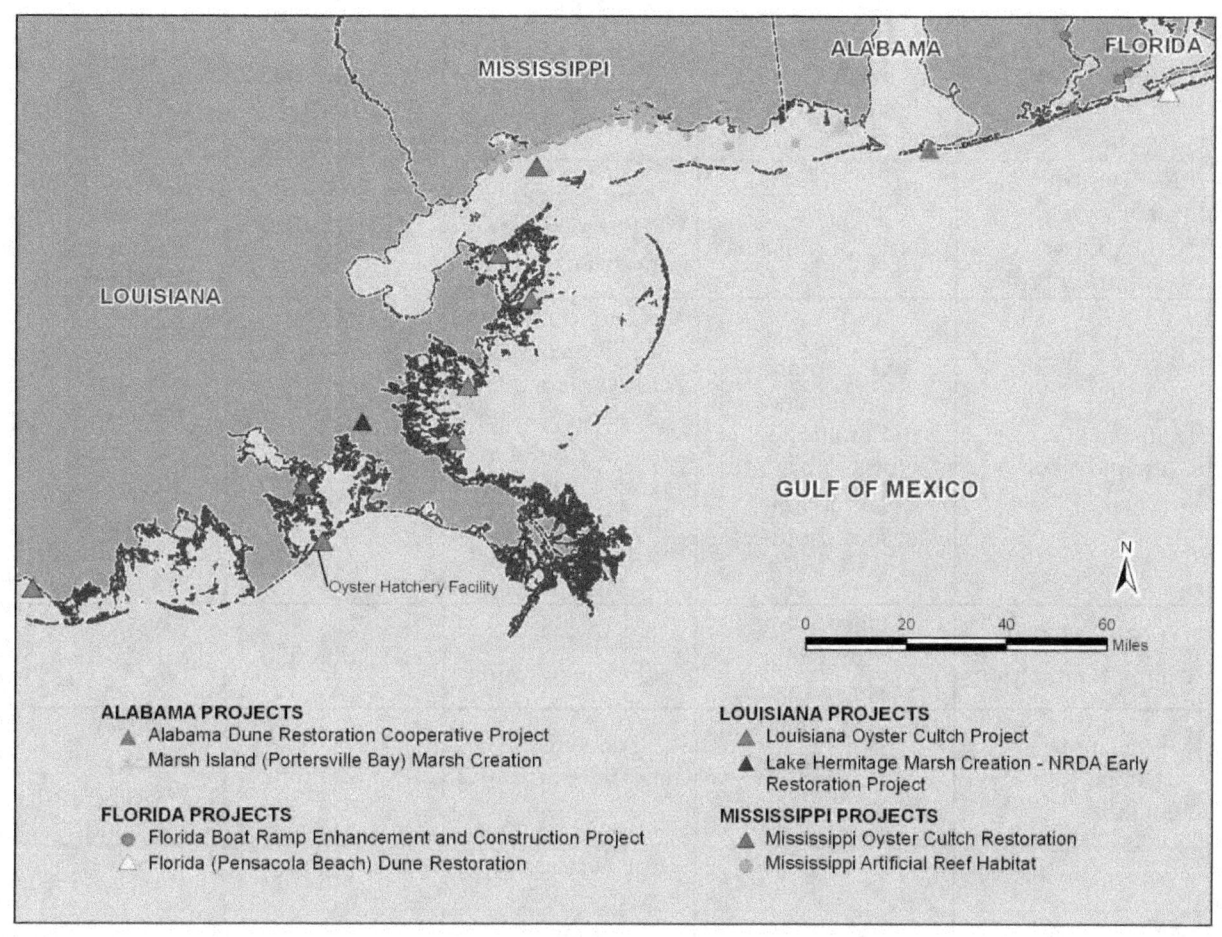

Figure ES-2: Location of Phase I Early Restoration projects.

Table ES-1. Phase I Early Restoration projects included in the selected action.

Project Title	Location (Parish/County and State)	Selected Restoration	Estimated Cost (including potential contingencies)[3]	Resources Benefitted
Lake Hermitage Marsh Creation – NRDA Early Restoration Project	Plaquemines Parish, Louisiana	Approximately 104 acres of marsh creation	$14,400,000	Brackish Marsh in the Barataria Hydrologic Basin
Louisiana Oyster Cultch Project	St. Bernard, Plaquemines, Lafourche, Jefferson, and Terrebonne Parishes, Louisiana	Approximately 850 acres of cultch placement on public oyster seed grounds; construction of improvements to an existing oyster hatchery	$15,582,600	Oysters in Coastal Louisiana
Mississippi Oyster Cultch Restoration	Hancock and Harrison Counties, Mississippi	1,430 acres of cultch restoration	$11,000,000	Oysters in Mississippi Sound
Mississippi Artificial Reef Habitat	Hancock, Harrison, and Jackson Counties, Mississippi	100 acres of nearshore artificial reef	$2,600,000	Nearshore Habitat in Mississippi Sound
Marsh Island (Portersville Bay) Marsh Creation	Mobile County, Alabama	protecting 24 existing acres of salt marsh; creating 50 acres of salt marsh; 5,000 linear feet of tidal creeks	$11,280,000	Coastal Salt Marsh in Alabama
Alabama Dune Restoration Cooperative Project	Baldwin County, Alabama	55 acres of primary dune habitat	$1,480,000	Coastal Dune and Beach Mouse Habitat in Alabama
Florida Boat Ramp Enhancement and Construction Project	Escambia County, Florida	Four boat ramp facilities	$5,067,255	Human Use in Escambia County, FL
Florida (Pensacola Beach) Dune Restoration	Escambia County, Florida	20 acres of coastal dune habitat	$644,487	Coastal Dune Habitat in Escambia County, FL

---

[3] Estimated costs for some of the projects were updated from those provided in the DERP/EA. Actual costs may differ depending on future contingencies, but will not exceed the amount shown without further agreement between the Trustees and BP.

Table of Contents

CHAPTER 1   BACKGROUND, PURPOSE AND NEED FOR PROPOSED ACTION

1.1   Introduction

On or about April 20, 2010, the mobile offshore drilling unit *Deepwater Horizon,* which was being used to drill a well for BP Exploration and Production, Inc. (BP) in the Macondo prospect (Mississippi Canyon 252 – MC252), experienced an explosion, leading to a fire and its subsequent sinking in the Gulf of Mexico. This incident resulted in discharges of oil and other substances from the rig and the submerged wellhead into the Gulf of Mexico. An estimated 5 million barrels (210 million gallons) of oil were subsequently released from the well over a period of approximately 3 months.[4] In addition, approximately 771,000 gallons of dispersants[5] were applied to the waters of the spill area in an attempt to minimize impacts from spilled oil.

The U.S. Coast Guard responded and directed federal efforts to contain and clean up the spill (hereafter referred to as "the Spill", which includes activities conducted in response to the spilled oil). At one point nearly 50,000 responders were involved in cleanup activities in open water, beach and marsh habitats. The magnitude of the Spill was unprecedented, causing impacts to coastal and oceanic ecosystems ranging from the deep ocean floor, through the oceanic water column, to the highly productive coastal habitats of the northern Gulf of Mexico, including estuaries, shorelines and coastal marsh. Affected resources include ecologically, recreationally, and commercially important species and their habitats in the Gulf of Mexico and along the coastal areas of Alabama, Florida, Louisiana, Mississippi, and Texas. These fish and wildlife species and their supporting habitats provide a number of important ecological and human use services.

1.2   Overview of the Oil Pollution Act and the National Environmental Policy Act

1.2.1   The Oil Pollution Act

The Oil Pollution Act Title 33 U.S.C. § 2701. *et seq.* (OPA), and the regulations for natural resource damage assessments (NRDA) under OPA, 15 C.F.R. Part 990, establish a liability regime for oil spills into navigable waters or adjacent shorelines that injure or are likely to injure natural resources and services that those resources provide to the ecosystem or humans. Pursuant to section 2706 of OPA, federal and state trustees for natural resources are authorized to (1) assess natural resource injuries resulting from a discharge of oil or the substantial threat of a discharge and response activities, and (2) develop and implement a plan for restoration of such injured resources.

---

[4] Oil Budget Team, OIL BUDGET CALCULATOR TECHNICAL DOCUMENTATION (November 23, 2010).

[5] Dispersants do not remove oil from the ocean. Rather, they are used to help break large globs of oil into smaller droplets that can be more readily dissolved into the water column.

The federal trustees are designated pursuant to the National Contingency Plan, 40 C.F.R. Section § 300.600 and Executive Order 12777. The following federal agencies are designated natural resources trustees under OPA and are currently acting as trustees for the Spill[6]:

- the United States Department of the Interior (DOI), as represented by the National Park Service, United States Fish and Wildlife Service, and Bureau of Land Management;
- the National Oceanic and Atmospheric Administration (NOAA), on behalf of the United States Department of Commerce.

State trustees are designated by the Governors of each state pursuant to the National Contingency Plan, 40 C.F.R. Section § 300.605. The following state agencies are designated natural resources trustees under OPA and are currently acting as trustees for the Spill:

- the State of Louisiana's Coastal Protection and Restoration Authority, Oil Spill Coordinator's Office, Department of Environmental Quality, Department of Wildlife and Fisheries and Department of Natural Resources;
- the State of Mississippi's Department of Environmental Quality;
- the State of Alabama's Department of Conservation and Natural Resources and Geological Survey of Alabama;
- the State of Florida's Department of Environmental Protection and Fish and Wildlife Conservation Commission; and
- for the State of Texas: Texas Parks and Wildlife Department, Texas General Land Office and Texas Commission on Environmental Quality.

Collectively, these federal and state entities are referred to as the "Trustees" throughout this document. In addition to acting as trustees for this incident under OPA, the States of Louisiana, Mississippi, Alabama, Florida and Texas are also acting pursuant to their applicable state laws and authorities, including:

- the Louisiana Oil Spill Prevention and Response Act of 1991, La. R.S. 30:2451 *et seq.*, and accompanying regulations, La. Admin. Code 43:101 *et seq.*;
- the Texas Oil Spill Prevention and Response Act, Tex. Nat. Res. Code, Chapter 40.01 *et seq*;
- the Florida Pollutant Discharge Prevention and Removal Act, Fla. Statutes Section 376.011 *et seq.*;
- the Mississippi Air and Water Pollution Control Law, Miss. Code Ann. §§ 49-17-1 through 49-17-43; and
- Alabama Code §§ 9-2-1 *et seq.* and 9-4-1 *et seq.*

Pursuant to OPA, federal and state agencies, Indian tribes and foreign governments may act as trustees on behalf of the public to assess the injuries and plan for restoration to compensate for those injuries. OPA further instructs the designated trustees to develop and implement a plan for

---

[6] The Department of Defense ("DOD") is also a trustee of natural resources associated with DOD-managed land on the Gulf Coast, which is included in the ongoing NRDA, but DOD is not a signatory of the Framework Agreement nor a participant in this Phase 1 Early Restoration Plan.

the restoration, rehabilitation, replacement, or acquisition of the equivalent of the injured natural resources under their trusteeship (hereafter collectively referred to as "restoration"). OPA defines "natural resources" to include land, fish, wildlife, biota, air, water sources, and other such resources belonging to, managed by, held in trust by, appertaining to, or otherwise controlled by the United States, any State or local government or Indian tribe, or any foreign government. This Phase I Early Restoration Plan (ERP) and Environmental Assessment (EA) (collectively referred to as the ERP/EA) was prepared jointly by the Trustees.

Natural resource services are the ecological and human use services that natural resources provide. Examples of ecological services include biological diversity, nutrient cycling, food production for other species, habitat provision, and other services that natural resources provide for each other. Human use services include activities that make 'direct' use of natural resources (e.g., boating, nature photography, education, fishing, swimming, hiking, etc.) as well as the value the public holds for natural resources independent of their own use of such resources (e.g., existence value, bequest value, etc.). For the purposes of this document the term "natural resource services" shall include these ecological and human use services.

1.2.2   The National Environmental Policy Act (NEPA)

NEPA, 42 U.S.C. § 4321, *et seq.* and its implementing regulations at 40 C.F.R. Parts 1500-1508 set forth a process of impact analysis and public review for federal agency actions, including restoration actions. NEPA provides a mandate and a framework for federal agencies to consider all reasonably foreseeable environmental effects of their proposed actions and to inform and involve the public in their environmental analysis and decision-making process.

Actions undertaken by federal trustees to restore natural resources or services under OPA and other federal laws are subject to NEPA, 42 U.S.C. § 4321 *et seq.*, and the regulations guiding its implementation at 40 C.F.R. Part 1500.[7] NEPA and its implementing regulations outline the responsibilities of federal agencies under NEPA, including the preparation of environmental documentation. In general, federal agencies contemplating implementation of a major federal action must produce an environmental impact statement (EIS) if the action is expected to have significant impacts on the quality of the human environment. When it is uncertain whether a contemplated action is likely to have significant impacts, federal agencies prepare an environmental assessment (EA) to evaluate the need for an EIS. If the EA demonstrates that the proposed action will not significantly impact the quality of the human environment, the federal agencies issue a Finding of No Significant Impact (FONSI), which satisfies the requirements of NEPA, and no EIS is required. If a FONSI cannot be made, then an EIS is required.

The Trustees prepared this ERP/EA in accordance with OPA NRDA regulations (see 15 C.F.R § 990.23) and NEPA requirements, which both require public involvement in the decision-making process. This ERP/EA presents information to the public regarding the affected environment, NRDA restoration planning, and actions designed to help address natural resource injuries and

---

[7] NEPA imposes legal requirements on federal trustees only.

lost human use of injured natural resources caused by the Spill. Restoration projects go beyond cleanup activities by restoring[8] injured natural resources or lost services.

The Phase I restoration alternative selected by the Trustees (see Chapter 3) is comprised of eight restoration projects. As discussed in Chapter 4, each project has been analyzed separately under NEPA because each project has independent utility. In accordance with NEPA and its implementing regulations, this ERP/EA summarizes the current environmental setting, describes the purpose and need for restoration, identifies restoration alternatives considered for injuries, assesses their applicability and potential environmental consequences, and summarizes the opportunity afforded for public participation in the process of making the Phase I early restoration plan decisions. This information has been used to make a threshold determination as to whether preparation of an EIS is required prior to selecting the final Phase I early restoration actions.

1.2.3 Compliance with other Applicable Authorities

In addition to the requirements of OPA and NEPA, requirements of other laws may apply to the early restoration planning or early restoration implementation. The Trustees will ensure compliance with all applicable authorities for all early restoration projects. To assist the public with identifying other applicable authorities, the Trustees prepared a non-exclusive list of other potentially applicable federal authorities attached as Appendix D. Whether and the extent to which an authority applies to a particular project depends on the specific characteristics of a particular project. Consequently, not every authority listed in Appendix D would apply to every project. In addition, state trustees will ensure compliance with applicable authorities in their individual states.

1.3   Natural Resource Damage Assessment Restoration Planning

Restoration activities are intended to restore or replace habitats, species, and services to their baseline condition (primary restoration), and to compensate the public for interim losses from the time natural resources are injured until they are restored or replaced to achieve baseline conditions (compensatory restoration). To meet these goals, the restoration activities need to produce benefits that are related, or have a nexus, to natural resources injured and associated service losses resulting from the oil spill, associated response or clean-up activities.

> **Restoration Terms Defined**
>
> Restoration: Any action that restores, rehabilitates, replaces, or acquires the equivalent of the injured natural resources.
>
> Primary Restoration: Any action that replaces or restores injured natural resources and services to their baseline condition.
>
> Compensatory restoration: Any action that replaces or restores the natural resource injuries and services lost from the date of injury until recovery to baseline conditions occurs.

NRDA restoration planning is designed to evaluate potential injuries to natural resources and natural resource services; to use that information to determine whether and to what extent restoration is needed; to identify potential restoration actions to address that need; and to provide

---

[8] For the purposes of this document, "restoring" or "restoration" includes any action that restores, rehabilitates, replaces, or acquires the equivalent of the injured natural resources or lost services.

the public with an opportunity to review and comment on the proposed restoration alternatives. Restoration planning has two basic components: (1) injury assessment and (2) restoration selection.

The goal of injury assessment is to determine the nature and extent of injuries to natural resources and services. The goal of restoration planning is to evaluate the need for and type of restoration required based on the injury assessment. Ultimately, trustees identify proposed restoration alternatives expected to compensate the public for losses of natural resources and services resulting from the spill.

Given its expansive geographic scale and complexity, the *Deepwater Horizon* NRDA may continue for years. In response to this extraordinary event, the Trustees initiated the restoration and planning efforts described below, even while damage assessment activities continue.

The early restoration projects selected in this ERP/EA are not intended to fully compensate the public for injuries caused by the Spill. Additional restoration actions will be required.

Emergency Restoration
Under OPA, trustees may take emergency restoration actions before completing the NRDA process in order to minimize continuing, or prevent additional, injury as long as the actions are feasible and the cost of the actions are reasonable.

The Trustees collectively implemented three emergency restoration projects as part of the Spill, addressing submerged aquatic vegetation, waterfowl, and sea turtles. The submerged aquatic vegetation project was implemented to prevent additional injury by restoring submerged aquatic vegetation beds damaged by propeller scarring and other response vessel impacts. The waterfowl habitat enhancements project provided alternative wetland habitat in Mississippi for waterfowl and shorebirds that might otherwise winter in oil-affected habitats. The sea turtle project was completed to improve the nesting and hatching success of endangered sea turtles on the Texas coast, including Padre Island National Seashore. Some Trustees also implemented additional response and emergency restoration actions independent of the other Trustees.

Gulf Spill Restoration Planning Programmatic Environmental Impact Statement (PEIS)
The Trustees are preparing a draft programmatic environmental impact statement (DPEIS) to address environmental impacts from and to facilitate the selection of restoration alternatives. Public input from scoping conducted as part of that process, and similar exercises conducted by individual Trustees, will also be considered in the development of early restoration plans (see Section 1.5 below). The DPEIS will assist the Trustees in making informed decisions regarding the selection and implementation of a range of restoration types that could be used to compensate the public and the environment for the loss of natural resources and services from the Spill. The Notice of Intent initiating this effort can be viewed at:
http://www.gulfspillrestoration.noaa.gov/wp-content/uploads/2011/02/PEIS-NOI_signed.pdf.

Early Restoration
On April 21, 2011, the Trustees entered into an agreement whereby BP is to provide $1 billion toward early restoration projects in the Gulf of Mexico to address injuries to natural resources

caused by the Spill. As described below, this early restoration agreement, entitled "Framework for Early Restoration Addressing Injuries Resulting from the *Deepwater Horizon* Oil Spill" (Framework Agreement)[9], represents a preliminary, initial step toward the restoration of injured natural resources. The Framework Agreement is intended to facilitate and expedite restoration in the Gulf in advance of the completion of the natural resource damage assessment process. The Framework Agreement provides a mechanism through which the Trustees and BP can work together "to commence implementation of early restoration projects that will provide meaningful benefits to accelerate restoration in the Gulf as quickly as practicable" prior to completion of the natural resource damage assessment process or full resolution of the Trustees' natural resource damage claims.

This ERP/EA addresses OPA and NEPA requirements for implementing Phase I early restoration projects. It includes a discussion of the alternative project proposals considered for Phase I and NEPA analyses for each of the selected projects. It is important to note that this ERP/EA is not intended to quantify the extent of restoration needed to satisfy claims under applicable law against the responsible parties; rather, the early restoration projects described herein are intended to accelerate meaningful restoration in the Gulf.

The ERP/EA also identifies the restoration benefits estimated to be provided by each project (referred to as "Offsets"). The term "Offsets" shall have the same meaning as provided in the Framework Agreement. Pursuant to the Framework Agreement, the Offsets were estimated using metrics that reflect natural resources and/or services expected to result from each project. At the end of the NRDA process, the Trustees will credit the Offsets identified for these early restoration projects against the total injury for the Spill. Further restoration will still be required to fully compensate the public for natural resource losses from the Spill.

For efficiency, the Trustees decided to evaluate each early restoration project in a single restoration plan. Consequently, the Draft Phase I ERP/EA included an evaluation of a no action alternative (Alternative A) and an evaluation of each proposed early restoration project (Alternative B). Under Alternative A (No Action – Natural Recovery), the Trustees would not implement any early restoration projects. Selecting this alternative would not have precluded analysis and implementation of different restoration activities at a later date. The selected alternative (Alternative B: Phase I Early Restoration Projects) describes eight separate projects that the Trustees concluded meet the evaluation criteria in Section 1.6 after considering public comment on the Draft Phase I ERP/EA. It is important to note that the projects in this ERP/EA represent only the first phase of the early restoration process. The Trustees continue to evaluate projects already submitted for consideration, as well as any new projects as they are received with the intent of proposing additional projects for the early restoration process.

In pursuing early restoration options, the Trustees are also mindful of other Gulf of Mexico restoration reports and related efforts, such as those by the Gulf Coast Ecosystem Restoration Task Force (GCERTF, 2011), Mabus (2010), Brown et al. (2011), NRCS (2011), Peterson et al. (2011) and others, including restoration planning efforts being undertaken by individual Trustees, such as Louisiana's Coastal Master Plan and Annual Plan updates and the Mississippi Coastal Improvements Plan (USACE, 2009).

---

[9] http://www.restorethegulf.gov/sites/default/files/documents/pdf/framework-for-early-restoration-04212011.pdf.

1.4   Purpose and Need for Early Restoration

The Phase I early restoration projects selected by the Trustees in this plan are designed to accelerate meaningful restoration in the Gulf and compensate the public for lost use of natural resources prior to completion of the full damage assessment. The projects are not intended to, and do not fully, address all injuries caused by the Spill.

1.5   Restoration Project Solicitation

Public input is an integral part of NEPA, OPA and the Spill restoration planning effort. Public review allows the public to consider and provide direct input to the Trustees on proposed restoration plans and alternatives and ensures that the Trustees can consider relevant information and concerns of the public prior to making final decisions on proposed actions.

Following the Spill, the Trustees established websites to provide the public information about injury and restoration processes.[10] A Notice of Intent to Conduct Restoration Planning for the *Deepwater Horizon* oil spill (Notice) was published in the Federal Register on October 1, 2010 and announced publicly by the Trustees. Pursuant to 15 C.F.R. § 990.44, the Notice announced that the Trustees determined to proceed with restoration planning to fully evaluate, assess, quantify, and develop plans for restoring, replacing, or acquiring the equivalent of natural resources injured and losses resulting from the Spill. Public solicitation of restoration projects has been on-going since publication of the Notice. The Trustees invited the public to participate in restoration planning for the Spill in accordance with 15 C.F.R. § 990.14(d) and State authorities, including hosting public meetings held across all the Gulf States during October, November and December 2010:

- October 12: Galveston, Texas
- October 25: Thibodaux, Louisiana
- October 26: Harahan, Louisiana
- October 27: New Iberia, Louisiana
- October 28: Chalmette, Louisiana
- November 11: Spanish Fort, Alabama
- November 18: New Orleans, Louisiana
- November 22: Long Beach, Mississippi
- November 30: Fort Walton Beach, Florida
- December 3: Tallahassee, Florida

These public meetings provided an opportunity for people to gain knowledge of the restoration process by speaking one-on-one with experts or asking questions in a town hall setting.

---

[10] See, www.fws.gov/contaminants/DeepwaterHorizon/DH_NRDA.cfm; www.gulfspillrestoration.noaa.gov; losco-dwh.com; www.dep.state.fl.us/deepwaterhorizon; www.mdeqnrda.com; http://www.tpwd.state.tx.us/landwater/water/environconcerns/damage_assessment/deep_water_horizon.phtm; www.outdooralabama.com

More broadly, the Trustees actively solicited public input through a variety of mechanisms, including public meetings, electronic communication, and creation of a Trustee-wide public website and database to share information and receive public project submissions. Non-electronic (hardcopy) submittals to the Trustees were also included into this database, located at www.gulfspillrestoration.noaa.gov. Some Trustees also constructed other localized websites to convey and collect public project submissions or comments.

The Trustees also hosted public meetings related to the development of the DPEIS related to the Spill. Public meetings for the DPEIS were held in March and April 2011, in each of the five Gulf States and Washington, DC, as follows:

- March 16:   Pensacola, Florida
- March 17:   Panama City, Florida
- March 21:   Biloxi, Mississippi
- March 22:   Belle Chasse, Louisiana
- March 23:   Mobile, Alabama
- March 24:   Houma, Louisiana
- March 28:   Grand Isle, Louisiana
- March 29:   Morgan City, Louisiana
- March 30:   Port Arthur, Texas
- March 31:   Galveston, Texas
- April 6:   Washington, D.C.

While not part of the early restoration planning process, the DPEIS scoping meetings provided useful background information related to the public's concern and interests regarding restoration ideas. The Trustees took advantage of that input in Phase I early restoration plan development.

Following adoption of the Framework Agreement in April 2011, the Trustees invited the public to provide restoration project ideas specific to the early restoration process through a variety of mechanisms, including internet-accessible databases.[11] The Trustees received hundreds of proposals, all of which can be viewed at these web pages. The Trustees also hosted public meetings in each of the five Gulf States in 2011 to explicitly solicit early restoration ideas:

- June 20: New Orleans, Louisiana
- June 8: Spanish Fort, Alabama
- June 9: Corpus Christi, Texas
- June 17: Santa Rosa Beach, Florida
- July 7: Biloxi, Mississippi
- July 12: Pensacola, Florida

Finally, the Trustees have addressed and continue to address NRDA, the restoration planning process and potential restoration projects at other public meetings and venues and meet with many non-governmental organizations and other potential stakeholders. The Trustees continue to

---

[11] See, www.gulfspillrestoration.noaa.gov; losco-dwh.com; www.dep.state.fl.us/deepwaterhorizon; www.mdeqnrda.com.

solicit restoration ideas via the web[12] and continue to consider existing and new project proposals as part of the restoration planning process. Figure 1 depicts the general project solicitation and selection process. In summary, project selection is a step-wise process comprised of: (1) project solicitation; (2) project screening and identification; (3) negotiation; and (4) public review and comment, described more fully below.

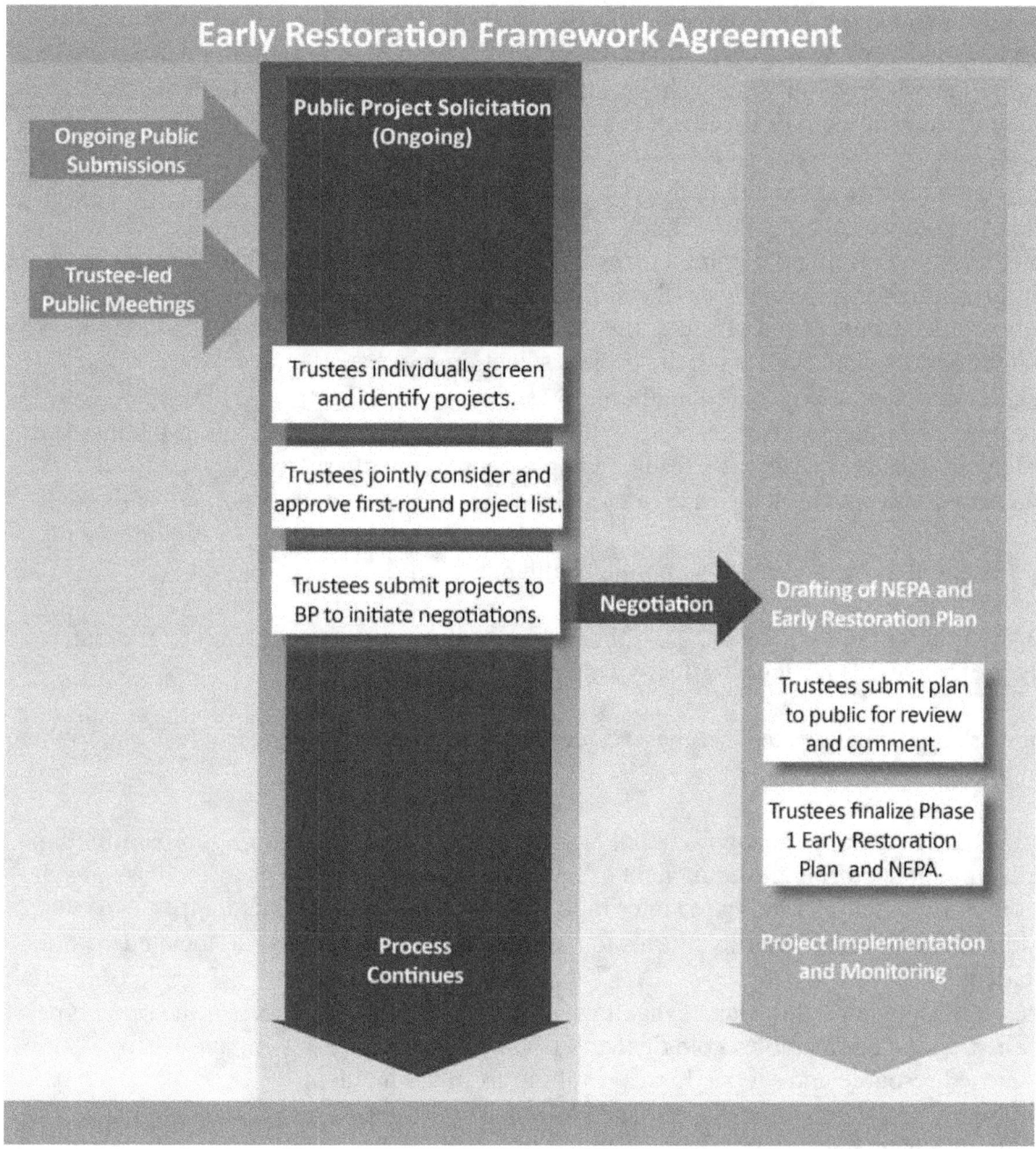

Figure 1. General Early Restoration project selection process.

---

[12] See, www.gulfspillrestoration.noaa.gov; losco-dwh.com; www.mdeqnrda.com; http://www.tpwd.state.tx.us/landwater/water/environconcerns/damage_assessment/deep_water_horizon.phtml www.outdooralbama.com, www.dep.state.fl.us/deepwaterhorizon.

1.6   Evaluation Criteria

In evaluating potential Phase I actions, the Trustees considered the broad suite of projects proposed through the project solicitation process. Proposals were evaluated based on criteria included in the OPA NRDA regulations, the Framework Agreement, as well as factors that are otherwise key components in planning or effecting early restoration, including those associated with other laws, regulations and programs. The OPA NRDA regulations (15 C.F.R. § 990.54) provide guidance concerning the evaluation and selection of projects designed to compensate the public for injuries caused by oil spills. These regulations require the Trustees to evaluate proposed restoration alternatives based on, at a minimum:

- The cost to carry out the alternative;
- The extent to which each alternative is expected to meet the Trustees' goals and objectives in returning the injured natural resources and services to baseline and/or compensating for interim losses (the ability of the restoration project to provide comparable resources and services, that is, the nexus between the project and the injury, is an important consideration in the project selection process);
- The likelihood of success of each alternative;
- The extent to which each alternative will prevent future injury as a result of the incident, and avoid collateral injury as a result of implementing the alternative;
- The extent to which each alternative benefits more than one natural resource and/or service; and
- The effect of each alternative on public health and safety.

Under OPA regulations (15 CFR 990.54), if the Trustees conclude that two or more alternatives are equally preferable, the most cost-effective alternative must be chosen.

The Framework Agreement states that the Trustees shall select projects for early restoration that meet all of the following criteria:

- Contribute to making the environment and the public whole by restoring, rehabilitating, replacing, or acquiring the equivalent of natural resources or services injured as a result of the Spill, or compensating for interim losses resulting from the incident;
- Address one or more specific injuries to natural resources or services associated with the incident;
- Seek to restore natural resources, habitats, or natural resource services of the same type, quality, and of comparable ecological and/or human-use value to compensate for identified resource and service losses resulting from the incident;
- Are not inconsistent with the anticipated long-term restoration needs and anticipated final restoration plan; and
- Are feasible and cost-effective.

Trustees also took into account several practical considerations that, while not legally mandated, are nonetheless useful and permissible to help screen the large number of potential qualifying projects. None of these practical considerations was used as a "litmus test"; rather, they were used as flexible, discretionary factors to supplement the decision criteria described above. For example, Trustees:

- took into account how quickly a given project is likely to begin producing environmental benefits;
- sought a diverse set of projects providing benefits to a broad array of potentially injured resources;
- focused on types of projects with which they have significant experience, allowing them to predict costs and likely success with a relatively high degree of confidence and making it easier to reach agreement with BP on the Offsets attributed to each project, as required by the Framework Agreement; and
- gave preference to projects that were closer to being ready to implement.

All of these discretionary factors are consistent with a key objective for pursuing early restoration: to secure tangible recovery of natural resources and natural resource services for the public's benefit while the longer-term process of fully assessing injury and damages is still underway.

In addition, OPA regulations (15 CFR 990.54) include specific guidance on the utilization of existing restoration projects and regional restoration plans (e.g., Louisiana Regional Restoration Plan, Region 2, NOAA et al., 2007a; Louisiana Regional Restoration Planning Program (RRP Program)[13]) to address natural resource injuries when appropriate. Projects already developed under such plans, with engineering designs, cost analyses, partner coordination, and permit and NEPA requirements satisfied, could be implemented quickly, and are good candidates for consideration in the early restoration process.

1.7   The Early Restoration Project Selection Process

The process that resulted in the selected alternative presented in this ERP/EA was developed by the Trustees to be responsive to the purpose and need for conducting early restoration. The Trustees acted promptly to identify project proposals that met the above criteria. Trustees evaluated proposals relative to the purpose and need for projects, potential impacts to the environment and selection criteria. Trustees identified preliminary lists of projects that were then brought to all of the Trustees for collective consideration and approval for the project negotiations with BP.

---

[13] Louisiana's RRP Program identifies the statewide Program structure, defines those trust resources and services in Louisiana that are likely to be or are anticipated to be injured (*i.e.,* at risk) by oil spill incidents, establishes a decision-making process, and sets forth criteria that are used to select restoration project(s) that may be implemented to restore the trust resources and services injured by a given spill. The RRP Program's Final Programmatic Environmental Impact Statement (FPEIS), which may be viewed in its entirety at http://www.losco.state.la.us/LOSCOuploads/RRPAR/la2395.pdf, is hereby incorporated by reference into this document.

1.8   Project Negotiation with BP

The OPA NRDA regulations require the Trustees to invite responsible parties to participate in the NRDA process. However, the authority and responsibility to assess natural resource injuries and losses and to define appropriate restoration plans rests solely with the Trustees. BP confirmed its interest in cooperatively participating in the NRDA process in 2010. The Framework Agreement evidences BP's willingness to support planning and implementing early restoration.

The process for selecting early restoration projects under the Framework Agreement began with project solicitation, development and evaluation by the Trustees as discussed above. The Trustees then engaged BP to determine whether an agreement in principle could be reached prior to inclusion of potential projects in a draft restoration plan. The Framework Agreement requires the Trustees and BP to agree on (1) the funding amount for a proposed project, and (2) Offsets. After the Trustees and BP reached an agreement in principle on these terms, these projects were combined into the Trustees' proposed alternative in the Phase I DERP/EA. However, the agreements can be finalized only after the public review process, described in more detail below.

1.9   Public Review and Comment

OPA, NEPA and the Framework Agreement require public input into the restoration process associated with the Spill. The Phase I DERP/EA served as a proposed restoration plan for Phase I of early restoration, environmental analyses of potential impacts of the projects, and the means used by the Trustees to seek public review and comment. The Trustees published the Phase I DERP/EA on December 15, 2011, and accepted comment on the draft for sixty (60) days following publication. A series of public meetings was held during that time in 2012 to facilitate the public review and comment:

- January 11: Fort Walton Beach, Florida
- January 12: Pensacola, Florida
- January 17: Gautier, Mississippi
- January 18: Gulfport, Mississippi
- January 19: Bay St. Louis, Mississippi
- January 23: Mobile, Alabama
- January 24: Gulf Shores, Alabama
- January 26: Galveston, Texas
- January 31: Houma, Louisiana
- February 1: Chalmette, Louisiana
- February 2: Belle Chasse, Louisiana
- February 7: Washington, D.C.

The Trustees considered comments on the DERP/EA prior to finalizing projects included in this Phase I ERP/EA. Summaries of comments received and Trustee responses are provided in Chapter 5 of this plan. Following publication of this ERP/EA, the Trustees will finalize agreements with BP regarding funding and offsets for the selected projects and proceed with implementation, subject to any remaining actions needed to comply with applicable state and federal laws.

1.10   Administrative Record

Pursuant to 15 C.F.R. § 990.45, the Trustees opened a publicly available administrative record (AR) for natural resource damage assessment and restoration activities concurrently with the publication of the Notice of Intent to Conduct Restoration Planning. DOI is the lead federal Trustee for maintaining the administrative record, which can be found at http://www.doi.gov/deepwaterhorizon/adminrecord. Some of the state Trustees are also maintaining a state-specific AR (e.g., loscodwh.com/AdminRecord.aspx). Information about project implementation will be provided to the public through the AR and other outreach efforts, including http://www.gulfspillrestoration.noaa.gov.

CHAPTER 2   ENVIRONMENTAL SETTING – GULF OF MEXICO

2.1   Introduction

This chapter describes the general environment of the Gulf of Mexico (Gulf) that provides the setting for the resources or services expected to benefit from the restoration projects included in this Phase I ERP/EA. These are resources and services that, even at this early stage in the NRDA process, are known to be impacted as a result of the Spill. These impacts provide the nexus for the early restoration projects included in this Phase I ERP/EA. Gulf physical, ecological and socioeconomic resources are generally described in Chapter 2. Additional information on the environmental setting for each early restoration project is also included in Chapter 4, as appropriate to the environmental analysis presented for each project in this Phase I ERP/EA for purposes of NEPA.

2.2   Physical Environment

The Gulf ecosystem is made up of a complex, intricate array of interconnected natural resources. These natural resources provide a wide range of services to both the environment, itself, and to humans. The U.S. Gulf coastline extends across five states: Florida, Alabama, Mississippi, Louisiana and Texas. The overall watershed that drains into the Gulf extends over more than 50% of the continental United States (USGS and EPA, 2011 as cited in GCERTF, 2011). The Mississippi-Atchafalaya River Basin alone drains an estimated 40 percent of the continental United States (NOAA, 2011a as cited in GCERTF, 2011).

Coastal and marine environments of the Gulf of Mexico include the intertidal zone, continental shelf, continental slope, and abyssal plain. The intertidal zone (also referred to as the foreshore or littoral zone) extends from mean lower low water to mean higher high water, and an upland area inward of mean higher high water. The upland area is not distinctly defined for this ERP/EA, but could include any area in the Gulf coast region potentially affected by a restoration project.

The continental shelf of the Gulf is seaward of the intertidal zone to the perimeter of the continental land mass. It can be divided into the inner and outer shelf environments. The extent of the continental shelf (miles from shoreline) and maximum depth at the shelf break varies throughout the basin. The inner continental shelf extends from mean lower low tide and is characterized by generally shallow waters and a gentle slope of a few feet per mile. The outer continental shelf is the deeper part of the shelf and extends to about a 650-foot depth contour.

Extending from the edge of the shelf to the abyssal plain, the outer continental slope is a steep area with diverse geomorphic features (canyons, troughs, and salt structures). The base of the slope in the Gulf occurs at a depth of about 9,000 feet. The Sigsbee Deep, located within the Sigsbee Abyssal Plain in the southwestern part of the basin, is the deepest region of the Gulf with a maximum depth ranging from about 12,000 to 14,000 feet (Figure 2).

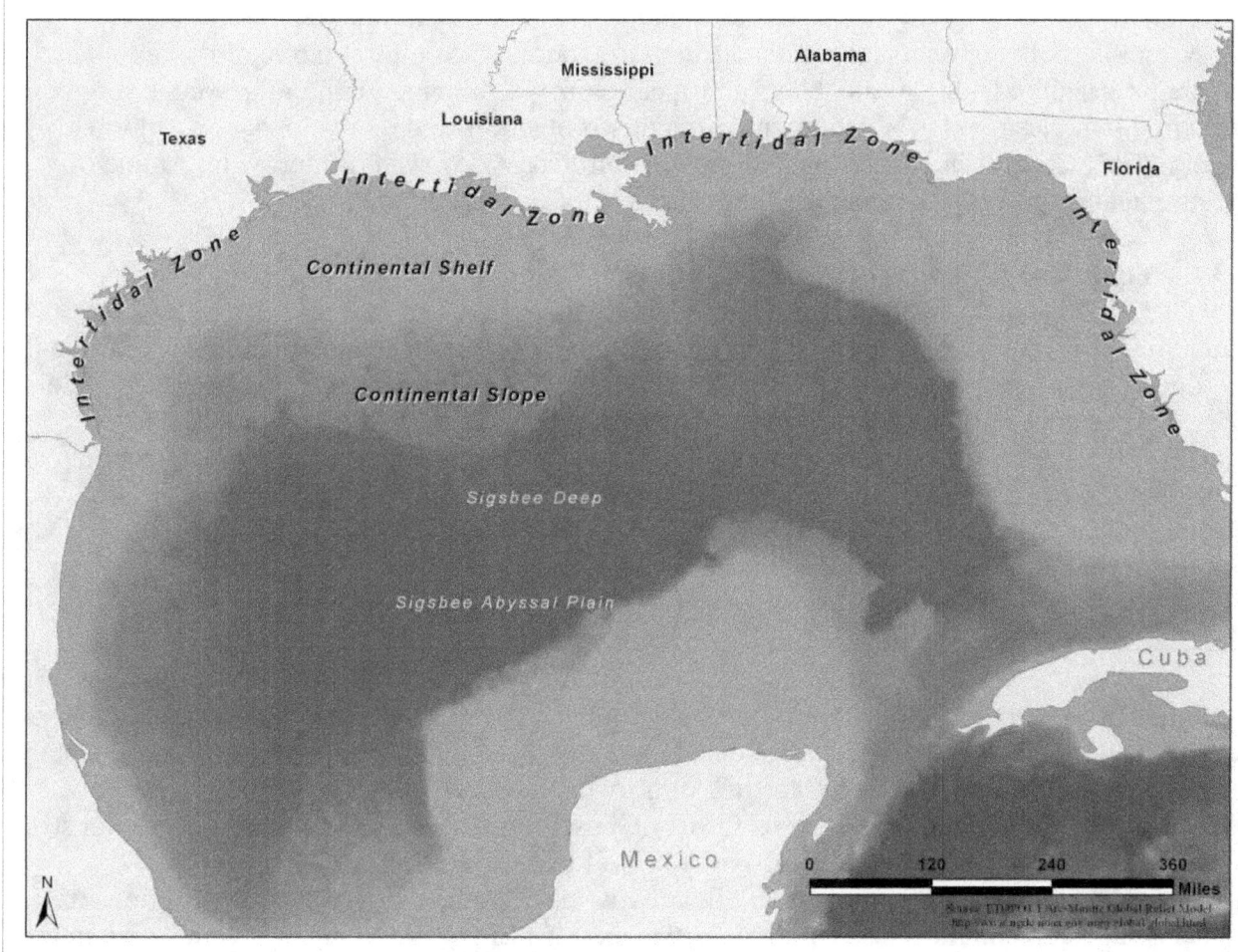

Figure 2. Gulf of Mexico.

2.3   Ecological Environment

The Gulf supports biologically diverse marine habitats and species, including planktonic communities, bottom-dwelling organisms, deepwater corals, sponges, fish, birds, terrestrial and marine mammals, and other species and communities. The Gulf is also home to a number of coastal, marine, and freshwater fish and wildlife species listed as threatened or endangered, as well as several species of protected marine mammals.

The Gulf supports a variety of coastal and marine habitats, including wetlands, barrier islands, beaches, seagrass beds, and coral and oyster reefs. These interconnected habitats are essential for the diverse array of ecologically, commercially, and recreationally important species that occur in the Gulf. For example, intertidal wetlands and other nearshore habitats (which extend from Texas to Florida) provide foraging and nesting habitats for the numerous species of birds using the Mississippi Flyway, one of the most important migratory bird flyways in the world. These coastal areas also provide essential habitats for ecologically, commercially, and recreationally important species of fish and invertebrates.

Individually and collectively, these coastal and marine habitats are integral to the Gulf ecosystem, to both regional and national economies, and to the cultural fabric of the region and the nation. Healthy Gulf Coast habitats and species provide a range of natural resource services including fisheries, food production, infrastructure protection, and recreational opportunities. Healthy Gulf Coast habitats also help to protect Gulf Coast communities, providing a line of defense against powerful storms, flooding and long term sea level rise.

2.3.1 Threatened, Endangered, and Candidate Species

Numerous species throughout the Gulf of Mexico are listed as threatened or endangered through the Endangered Species Act of 1973 (ESA). These species are protected and as provided under ESA, federal consultations are required when environmental actions may affect these listed species. Listed species potentially present in project areas are noted in Appendix B. Specific consideration of potential impacts to these species from these early restoration projects are further discussed in Chapter 4. ESA consultation correspondence will be available in the Administrative Record.

2.3.2 Essential Fish Habitat

Essential fish habitat (EFH) encompasses waterbodies, habitats, and substrates necessary for federally and regional fishery management council managed fish to complete various life history stages such as breeding, spawning, feeding or growth and survival to maturity. To comply with requirements of the Magnuson-Stevens Fishery Conservation and Management Act, the Trustees obtained information on designated EFH in the Gulf of Mexico from NOAA at http://www.habitat.noaa.gov/protection/efh/newInv/index.html, and from text descriptions in Fishery Management Plans also available at that site. An EFH assessment was completed on the Phase I DERP by the National Marine Fisheries Service, which concluded that the proposed actions would not adversely affect EFH, and, overall, would likely benefit federally managed fishery species. Specific consideration of potential impacts to these essential habitats from proposed early restoration projects are further discussed in Chapter 4. EFH consultation correspondence will be available in the Administrative Record. Representative EFH categories are listed in Table 1.

Table 1. Representative Categories of Essential Fish Habitat Identified in the Fishery Management Plan Amendment of the Gulf of Mexico Fishery Management Council.[14]

Estuarine areas	Marine areas
Estuarine Emergent Wetlands Estuarine Scrub/Shrub Mangroves SAV Oyster Reef and Shell Banks Intertidal Flats Palustrine emergent and forested wetlands Mud/sand/shell/rock substrates Estuarine water column	Coral and coral reefs Non-vegetated bottoms Artificial Reefs Water Column Live/Hard Bottom SAV

2.4   Socioeconomic Environment

The Gulf of Mexico is among the nation's most valuable and important ecosystems. The Gulf Coast and its natural resources are key components of the U.S. economy, producing 30 percent of the nation's gross domestic product in 2009 (NOAA, 2011b as cited in GCERTF, 2011). The region provides more than 90 percent of the nation's offshore oil and natural gas production (USEIA as cited in GCERTF, 2011); 33 percent of the nation's seafood (NOAA 2010 as cited in GCERTF, 2011); 13 of the top 20 ports by tonnage in the United States in 2009 (U.S. Army Corps of Engineers [USACE], 2010 as cited in GCERTF, 2011); as well as regionally and nationally important tourism and recreational activities such as fishing, boating, beachcombing, and bird watching. These activities support more than 800,000 jobs (Mabus, 2010 as cited in GCERTF, 2011) across the region, providing a substantial economic input to Gulf communities and the nation. All of these industries depend on a healthy and resilient Gulf. The five U.S. Gulf Coast States, if considered an individual country, would rank seventh in global gross domestic product (NOAA, 2011b as cited in GCERTF, 2011).

2.5   Cultural Resources

The Northern Gulf of Mexico has a rich cultural heritage. Cultural resources are prehistoric, historic, or archaeological services that have cultural significance and can include shipwrecks, historical buildings, monuments, and burial grounds. Cultural resources include historic properties listed in, or eligible for listing in the National Register of Historic Places (36 CFR §60[a-d]). The National Historic Preservation Act of 1966, as amended (NHPA; 16 U.S.C. §470(f)), defines an historic property as "any prehistoric or historic district, site, building, structure, or object included in, or eligible for inclusion on the National Register [of Historic Places]." This includes significant properties of traditional religious and/or cultural importance to Indian tribes.

Historic properties include built resources (bridges, buildings, piers, etc.), archaeological sites, and Traditional Cultural Properties, which are significant for their association with practices or

---

[14] EFH for species managed under the NMFS Billfish and Highly Migratory Species plans generally falls within marine and estuarine water column habitats designated by the Council.

beliefs of a living community that are both fundamental to that community's history and a piece of the community's cultural identity. Although often associated with Native American traditions, such properties also may be important for their significance to ethnic groups or communities. Historic properties also include submerged resources. Modern technology enables nautical archaeologists to recover data in areas previously inaccessible. The variety of shipping channels in the Gulf of Mexico encompasses colonial and modern-day trade routes and activities. In addition, armed conflicts from colonial times to the 1940s have left indelible marks on the Gulf Coast. Shipwrecks can range from seventeenth century Spanish galleons to World War II-era German U-boats. Small pirogues or canoes may provide data on Native American or local history. Maritime archaeology includes but is not limited to the study of wrecks; wrecks encompass airplane and boat debris.

Bridges, shell middens, harbors, and villages can be submerged as a result of changing coastlines and other climatic activity. Approximately 19,000 years ago, global sea level was approximately 120 meters lower than present. During this time, large expanses of what is now the outer continental shelf were exposed as dry land. Twelve thousand years ago, the earliest date prehistoric human populations are known to have been in the Gulf Coast region (Aten, 1983, as cited in MMS, 2007), sea level would have been approximately 45 meters lower than present day levels (CEI, 1982, as cited in MMS, 2007). The location of the shoreline 12,000 years ago is roughly approximated by the 45 meter bathymetric contour. The continental shelf shoreward of this contour would have potential for prehistoric sites dating subsequent to 12,000 years ago. Since known prehistoric sites on land usually occur in association with certain types of geographic features, prehistoric sites should be found in association with those same types of features now submerged and buried on the continental shelf.

Geographic features that have a high potential for associated prehistoric sites include barrier islands and back barrier embayments, river channels and associated floodplains, terraces, levees and point bars, and salt dome features. A review of previously identified archaeological work in the vicinity of a project is critical to determining the scope of the archaeological identification effort. Areas subjected to previous extensive archaeological investigations may not warrant additional fieldwork. All previous work should be evaluated in consultation with State Historic Preservation Office and, if involved, a Tribal Historic Preservation Officer for reliability and accuracy.

2.6   Socioeconomic and Environmental Justice

To the greatest extent practicable, federal agencies must "identify and address, as appropriate, disproportionately high and adverse human health or environmental effects of its programs, policies, and activities on minority populations and low-income populations." Executive Order 12898 (Feb. 11, 1994). The Council on Environmental Quality (CEQ) issued guidance directing federal agencies to analyze the environmental effects, including human health, economic, and social effects, of their proposed actions on minority and low-income communities when required by NEPA. CEQ, Environmental Justice: Guidance Under the National Environmental Policy Act, p. 25 (CEQ, 1997). CEQ defined members of minority populations to include: American Indian or Alaskan Native; Asian or Pacific Islander; Black, not of Hispanic origin; or Hispanic. Low income populations for this analysis were determined based on the U.S. Census Bureau 1999 poverty thresholds (USDOC, U.S. Census Bureau, 1999). Analyses in this ERP/EA comply with Executive Order 128898 and CEQ's guidance.

2.7   The *Deepwater Horizon* Oil Spill Natural Resource Damage Assessment (NRDA)

The Spill presents a complex threat to the interconnected organisms, habitats, and ecosystems of the Gulf of Mexico. Unprecedented volumes of oil and dispersants were released into the environment and were transported in deepwater areas, the water column, along the ocean's surface, through coastal and nearshore areas, and onto shorelines. Figure 3 illustrates some of the various types of resources and services being evaluated as part of the *Deepwater Horizon* NRDA and provides a sense of the scope of investigations being done to fully evaluate the impacts of oil, dispersants, and other response actions on natural resources and the Gulf ecosystem.

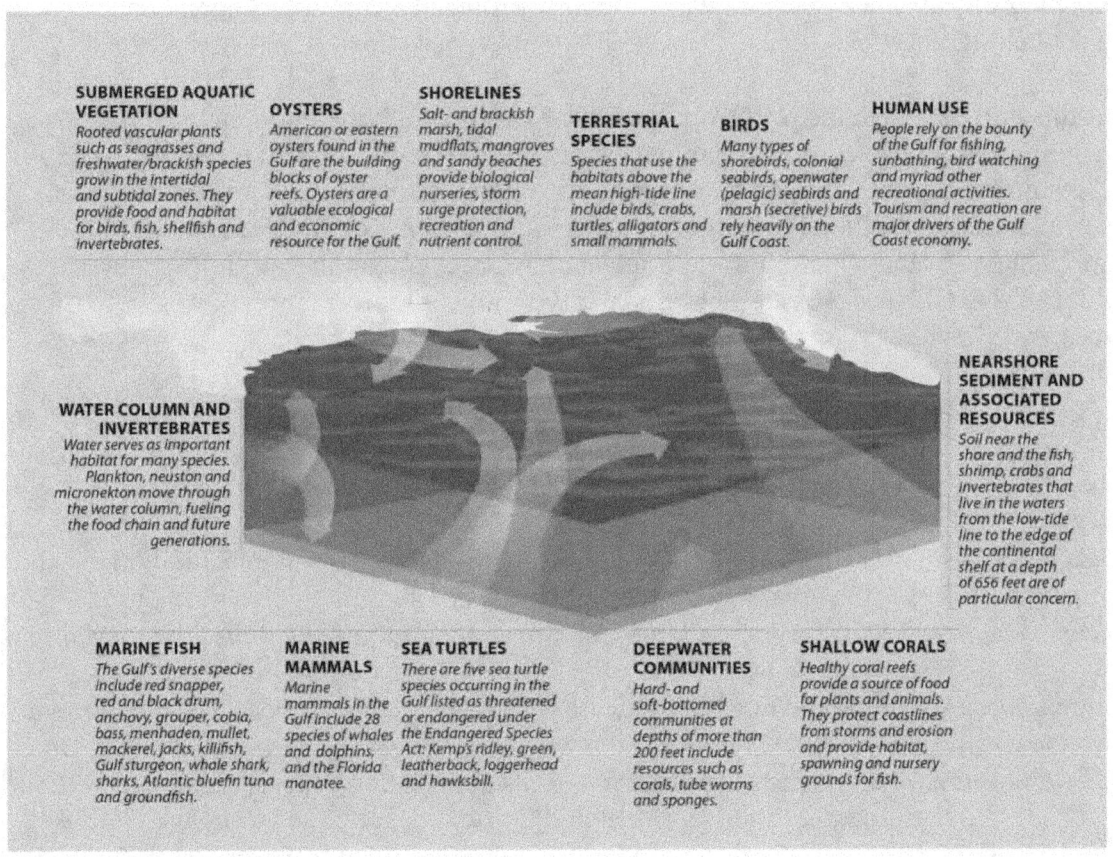

Figure 3. Gulf of Mexico resources potentially affected by the *Deepwater Horizon* Spill.

The *Deepwater Horizon* NRDA includes assessment and evaluation of potential injuries to a wide array of natural resources, from the deep ocean to the coastlines of the northern Gulf of Mexico. The injury assessment for the Spill is ongoing. Information continues to be collected to assess potential impacts to fish, shellfish, terrestrial and marine mammals, turtles, birds, and other sensitive resources as well as their habitats, including, but not limited to, wetlands, beaches, mudflats, bottom sediments, corals, and the water column. Lost human uses of these resources, such as recreational fishing, boating, hunting, and beachgoing, are also being assessed. Hundreds of scientists, economists, and restoration specialists have been and continue to be involved in these diverse NRDA activities.

Among the most readily observable impacts that have been a consequence of the Spill stem from the Gulf-wide response efforts aimed at reducing the short-term effects of oiling. These response efforts were undertaken at a massive scale, with nearly 50,000 responders active during the height of clean-up efforts. In addition, there were nearly 10,000 vessels involved in oil containment and removal, and millions of feet of absorbent and containment oil boom were deployed in an effort to reduce the amount of oil stranded along coastal shorelines. Although response efforts succeeded in reducing the amount of oil that was stranded on coastlines, these actions caused a number of unavoidable physical consequences on coastal resources, including smothering, trampling, removal, and disruptions in recreational use of beaches and waterways. Natural resource impacts associated with response actions have not fully been quantified, and some may be ongoing.

Even at this early stage in the NRDA process, and even though the nature and extent of natural resource injuries and losses are still being assessed, some of the adverse effects of the Spill on natural resources or services have been observed and/or reasonably inferred. Because this Phase I ERP/EA includes restoration projects with a nexus to injuries to coastal marsh, oysters, nearshore habitats, sandy shoreline and dune habitats, and human use of Gulf resources, the remainder of this Chapter provides additional environmental information pertinent to these resources.

2.7.1   Coastal Marshes

The Phase I ERP/EA includes two marsh restoration projects discussed in Chapters 3 and 4.

Oil made landfall on shorelines of Texas, Louisiana, Mississippi, Alabama and Florida. To date, preliminary estimates reflect that more than 4,000 miles of shoreline have been surveyed and oiling has been observed along more than 1,000 miles of shoreline, including in coastal marshes.

Coastal marshes are among the most biologically productive coastal areas in the continental U.S. and provide a critical ecological connection between coastal and open water habitats. Brackish and salt water marshes are found at the margins of estuaries, along barrier islands, and in tidal deltas. These marshes trap and filter sediment and nutrients, moderate freshwater inflows, provide habitat for migratory and resident wildlife, and provide nursery areas for shellfish and fish.

Wetlands along the Gulf coast include salt and brackish marsh environments. Salt marsh habitat is defined by clearly distinguishable zonation between low, middle, and high marsh elevation. The low marsh area is colonized primarily by *Spartina alterniflora* (smooth cordgrass) along with other small cordgrass and succulent species that are characteristic of this zone. Once the low marsh area is developed, sedimentation and debris build up and contribute to the development of the middle and high marsh zones (Bertness 1999). Dominant high marsh vegetation consists of *Spartina patens* (saltmeadow cordgrass) and *Juncus romerianus* (black needlerush) (U.S.

Department of Agriculture [USDA] 2011). Species found in the high marsh zone are generally less tolerant of flooding and high salinity than plants in the low marsh. Larger, highly branched woody species including many species of forbs are found primarily in the high marsh.

Brackish marshes generally form along the upland edge of salt marshes where freshwater input dilutes the salinity, creating brackish conditions (i.e., 0.5-18 ppt salinity). This environment supports species that are intolerant of extremely high or low salinities as well as species that are restricted to brackish conditions. Plant diversity is higher in brackish marshes as compared to salt marshes due to lower salinity stress. Decomposition rate, net primary production, and organic matter accumulation are also generally increased in brackish marshes (Odum, 1988). Dominant vegetation often overlaps with the high marsh zone of salt marshes and includes *Spartina patens*, *Distichlis spicata*, and *Juncus romerianus*. Brackish marshes along the Gulf frequently support a wide variety of plants, including *Schoenoplectus californicus* (California bulrush), *Eleocharis cellulosa* (Gulf spikegrass), and *Solidago sempervirens* (seaside goldenrod). Both types of marsh habitat also harbor a variety of marine and terrestrial species that utilize the productive environment for shelter, foraging, and breeding.

Extensive oiling of intertidal marsh habitats as a consequence of the Spill has been observed and documented in the northern Gulf. Visible oiling has been documented by Shoreline Clean-up Assessment Teams (SCAT) and in NRDA studies that are ongoing. For example, in Louisiana, preliminary estimates of field data reflect over 400 miles of intertidal marsh coastline were observed to have some degree of oiling from the time oil was first released through October 22, 2010. SCAT surveys and on-going NRDA studies have also revealed observable impacts from response activities in marsh habitats, including from vessels, booms, and oil removal. The adverse impacts from the Spill are still being assessed by the Trustees.

2.7.2   Oysters

The Phase I ERP/EA includes two oyster restoration projects discussed in Chapters 3 and 4.

The American, or eastern oyster (*Crassostrea virginica*), is the primary oyster species found in the Gulf. This species typically lives in shallow, well-mixed estuaries, lagoons, and oceanic bays. American oysters in the Gulf are found at elevations ranging from about 1 foot above the mean low tide line to about 4 feet below the mean low tide line. Oysters are tolerant of a wide range of temperatures, salinities, and concentrations of suspended solids. Oysters in the Gulf live on hard substrate along the coast and shallow intertidal areas. They prefer to attach to other oysters, but have also been found attached to other hard substrate, including man-made materials. This species is also an important economic contributor to the Gulf's economy. In fact, the region leads the nation in the production of oysters (about 67% of the nation's total).

Oyster exposure to oil and dispersants could have occurred through a variety of ways, such as swimming (dermal contact), feeding, drinking, and breathing for early life stages (e.g., larvae) and through filtration (feeding) for adult life stages. Oil has the potential to impact spawning success. The Spill occurred during the peak spawning period for oysters. Once these species spawn, the early larval stages move with the currents near the surface of the water and are unable to actively avoid potential exposure to oil and dispersants. It is known that oyster spawning grounds were exposed to oil.

2.7.3   Nearshore Habitats

The Phase I ERP/EA includes one nearshore habitat (reef) restoration project discussed in Chapters 3 and 4.

Nearshore habitats include sandy bottom sediments as well as hard bottom habitats such as oyster reefs, mussel beds and shell hash mound systems, each with their own diverse group of associated fauna. The nearshore non-vegetated sediment of the Gulf of Mexico serves as a diverse and essential habitat for many organisms. Nearshore sediments are rich with worms and bacteria that feed on organic material in the sediments.

Oil and dispersants reaching the nearshore environment were predominantly transported in the upper reaches of the water column by wind and currents. There are several pathways for this surface oil to reach nearshore sediments. Oil droplets may be adsorbed onto marine non-living organic material or sediments and sink. Oil that arrived on shore may have mixed with sediment and washed back out with the tide, eventually settling to the bottom. This sinking oil creates a hazard to the wide variety of organisms that live in the nearshore environment, including grasses, fish, crabs, shrimp, and other invertebrates. Many of these animals forage in the sediments for food and are susceptible to oil through dermal contact, intake by respiration, and ingestion.

2.7.4   Sandy Shorelines & Dune Habitat

The Phase I ERP/EA includes two sandy shoreline and dune habitat restoration projects discussed in Chapters 3 and 4.

The Gulf of Mexico has hundreds of miles of sandy shoreline that are important both ecologically and economically. Beaches and barrier islands along the Gulf coast vary between geographic regions, based on their respective geologic formation. Coastal dunes are a critical beach habitat that supports a variety of plant and animal species. Dunes are wind-blown sand mounds that form just behind the beach face. Although the regulatory definition of primary and secondary dunes may vary among jurisdictions, primary dunes are the foremost structures and thus incur most of the saline and thermal stress from coastal physical processes. Vegetation diversity is generally lower on primary dunes due to these factors. Secondary dunes are older and more stable and support more diverse and larger vegetation such as shrubs and small trees. A swale typically forms in between primary and secondary dunes and often supports plant species more tolerant to water inundation because this area acts as a catch for water that breaches the primary dune. Typical dune plants along the Gulf include *Panicum amarum* (bitter panicgrass) and other beach grasses along with cordgrasses such as *Spartina patens* (saltmeadow cordgrass).

There was extensive oiling of sandy beaches in the northern Gulf. This oiling was readily observable and documented in media coverage, in aerial photography, and in SCAT records. For example, in Alabama, approximately 80 miles of beaches were exposed to *Deepwater Horizon* oil, including about 39 miles experiencing heavy to moderate oiling. Response efforts were necessary and undertaken to remove oil from beaches. These activities have resulted in beach areas being closed, or in disruptions in enjoyment and recreational use of these resources. Response efforts also physically impacted beaches, including associated dune habitats, as a result of effects from motorized vehicles, trampling, as well as removal of sand, vegetation, wrack, and shell, which are important biotic habitats. Continuous disturbance by response activities can prevent typical seaward expansion of dunes. Media coverage, aerial photography, SCAT records and other observational data include evidence of these physical impacts to beaches and associated dune habitats. Work to assess the full extent of these injuries is ongoing.

2.7.5 Human Use

The Phase I ERP/EA includes one human use project discussed in Chapters 3 and 4.

Humans rely on the natural resources of the Gulf. Outdoor recreationists make millions of trips per year to the Gulf. Fishing, boating, education, beachgoing, and bird watching are among the many of recreational activities undertaken by Gulf residents and visitors. Tourism and recreation are large contributors to the Gulf economy. The sand beaches of the northern Gulf coast are important recreational destinations and vital tourist attractions that fuel local economies. The Spill affected public use and enjoyment of many of the natural resources across the Gulf. For example, public beach use was disrupted during response activities.

CHAPTER 3   ALTERNATIVES, INCLUDING THE SELECTED ALTERNATIVE

Through the April 21, 2011 Framework Agreement, BP agreed to provide up to $1 billion toward early restoration projects in the Gulf of Mexico to address injuries to natural resources caused by the Spill. The Framework Agreement represented a preliminary step toward the restoration of injured natural resources, and is intended to accelerate meaningful restoration in the Gulf in advance of the completion of the assessment process. Below we describe two alternatives that the Trustees considered for Phase I early restoration: the No Action alternative and the alternative selected by the Trustees.

3.1   Alternative A: No Action – Natural Recovery

Under the No Action alternative, the Trustees would not implement the early restoration projects identified in the Phase I ERP/EA and would rely solely on natural recovery to restore natural resources and associated services until the NRDA and final restoration are complete. Choosing this alternative would not preclude analysis and implementation of different restoration activities at a later date. The No Action alternative was used as a basis for comparison of the effects from implementing the alternatives. The baseline for comparison of the alternatives is defined as the current condition and expected future condition in the absence of the project(s).

3.2   Alternative B: Selected Alternative – Phase I Early Restoration Projects

Following the intent of the Framework Agreement and public comment on the DERP/EA, the Trustees selected and intend to move forward with the early restoration projects included in Alternative B; the Marsh Island (Portersville Bay) Marsh Creation project is approved for completion of project design, NEPA analysis and work necessary to support application for permits. NEPA review for the Marsh Island (Portersville Bay) Marsh Creation project would be completed before any implementation occurs. The Trustees will now seek to finalize agreements for each project with BP (see Section 1.8) as soon as possible, consistent with the Framework Agreement. While the Selected Alternative constitutes a suite of projects, each project is viewed as an independent action from the others and will proceed independently and in such time and manner as is appropriate to that project.

Restoration actions selected under the Framework Agreement are not intended to provide the full extent of restoration needed to satisfy claims against BP. The Trustees anticipate that additional projects will be proposed and approved in the early restoration process as it continues. Furthermore, after injury assessment activities are complete, there will be additional opportunities for consideration of projects as the NRDA restoration planning process moves forward.

Table 2 provides a brief overview of the projects selected for this ERP/EA. Projects are identified in geographic order, moving from West to East. Figure 4 illustrates project locations.

Table 2. Phase I Early Restoration Projects.

Project Title	Location (Parish/County and State)	Selected Restoration	Estimated Cost (including potential contingencies)[15]	Resources Benefitted
Lake Hermitage Marsh Creation – NRDA Early Restoration Project	Plaquemines Parish, Louisiana	Approximately 104 acres of marsh creation	$14,400,000	Brackish Marsh in the Barataria Hydrologic Basin
Louisiana Oyster Cultch Project	St. Bernard, Plaquemines, Lafourche, Jefferson, and Terrebonne Parishes, Louisiana	Approximately 850 acres of cultch placement on public oyster seed grounds; construction of improvements to an existing oyster hatchery	$15,582,600	Oysters in Coastal Louisiana
Mississippi Oyster Cultch Restoration	Hancock and Harrison Counties, Mississippi	1,430 acres of cultch restoration	$11,000,000	Oysters in Mississippi Sound
Mississippi Artificial Reef Habitat	Hancock, Harrison, and Jackson Counties, Mississippi	100 acres of nearshore artificial reef	$2,600,000	Nearshore Habitat in Mississippi Sound
Marsh Island (Portersville Bay) Marsh Creation	Mobile County, Alabama	protecting 24 existing acres of salt marsh; creating 50 acres of salt marsh; 5,000 linear feet of tidal creeks	$11,280,000	Coastal Salt Marsh in Alabama
Alabama Dune Restoration Cooperative Project	Baldwin County, Alabama	55 acres of primary dune habitat	$1,480,000	Coastal Dune and Beach Mouse Habitat in Alabama
Florida Boat Ramp Enhancement and Construction Project	Escambia County, Florida	Four boat ramp facilities	$5,067,255	Human Use in Escambia County, FL
Florida (Pensacola Beach) Dune Restoration	Escambia County, Florida	20 acres of coastal dune habitat	$644,487	Coastal Dune Habitat in Escambia County, FL

---

[15] Estimated costs for some of the projects were updated from those provided in the DERP/EA. Actual costs may differ depending on future contingencies, but will not exceed the amount shown without further agreement between the Trustees and BP.

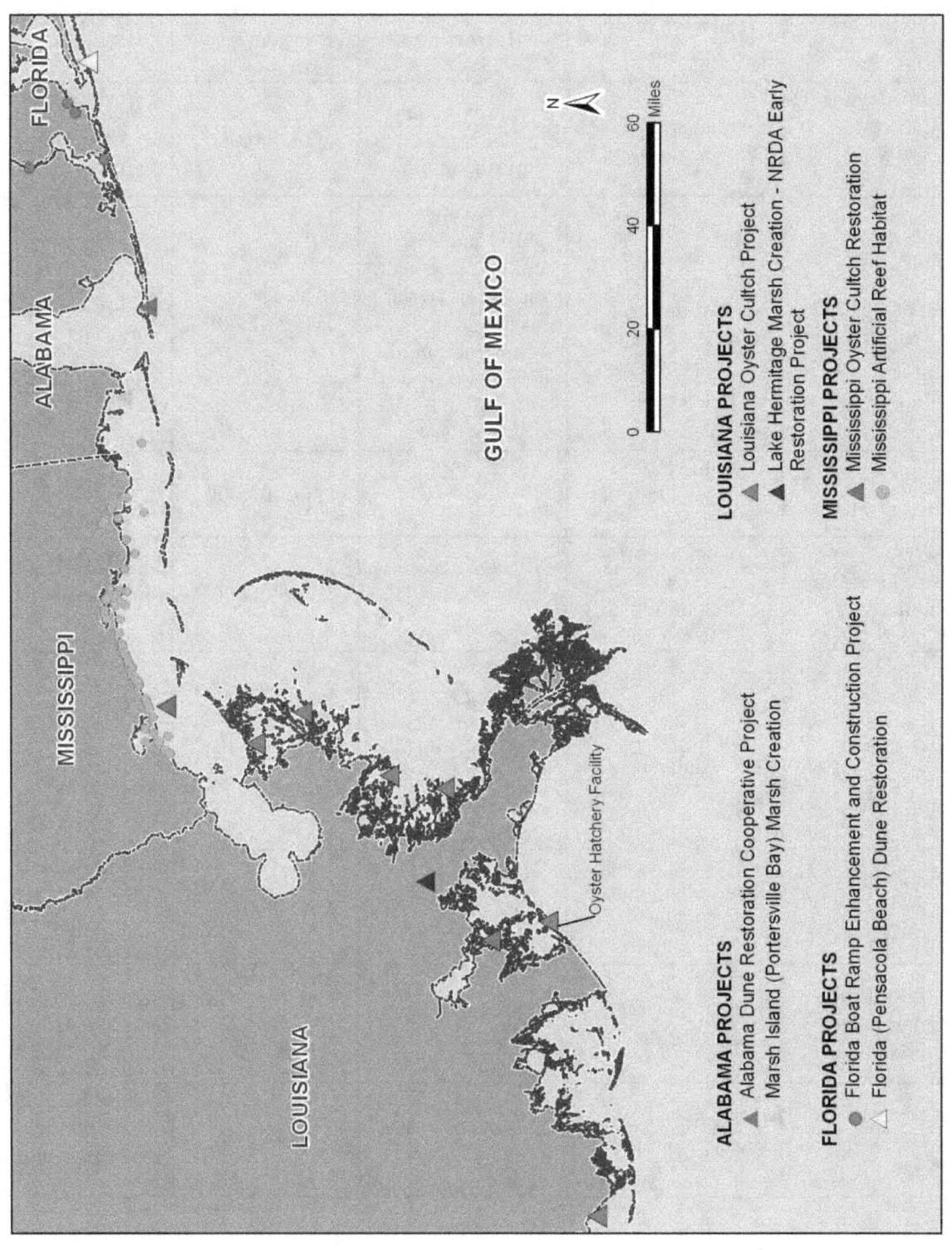

Figure 4. Location of Phase I Early Restoration projects.

3.2.1 Offsets Estimation Methodology for Projects

The Trustees used three primary methods to estimate Offsets for early restoration projects: Habitat Equivalency Analysis (HEA), Resource Equivalency Analysis (REA), and monetized estimates of project benefits. An overview of the Trustees' approach to estimating Offsets is outlined for each early restoration project.

HEA and REA are methods commonly used in natural resource damage assessments. HEA is used to quantify changes in ecological services on a habitat basis (e.g., units of marsh habitat) whereas REA is used to quantify ecological services[16] in resource specific units (e.g., oyster biomass). When HEA or REA is used to estimate restoration credits, anticipated ecological benefits resulting from the restoration action often are expressed in units that reflect the present (current) value of ecological benefits over a project's lifespan. For purposes of the early restoration projects included herein, the Trustees expressed HEA-estimated habitat benefits as "discounted service acre years" or DSAYs of the specific habitat types to be restored. For example, the Trustees estimated the present value of Offsets associated with early restoration projects focused on primary dune restoration in terms of primary dune DSAYs.

REA-estimated benefits are expressed in resource-specific units, rather than on a habitat basis. For example, the Trustees estimated the present value of ecological credits associated with early restoration projects focused on oyster cultch placement in terms of discounted kilogram years (DKg-Y) of oyster productivity.

The Trustees considered a variety of project-specific factors when applying HEA and REA methods to estimate the ecological benefits of restoration projects, including, but not limited to:

- the time at which ecological services from a restoration project begins to accrue;
- the rate of ecological service accrual over time;
- the time period over which ecological services will be provided;
- the quantity and quality of ecological services provided by the restored habitat or resource relative to those not affected by the Spill; and
- the size of the restoration action.

The benefits of a restoration project can also be monetized, or expressed in terms of a dollar value rather than in terms of ecological credits. Monetized benefits can be expressed in terms of the present value of project implementation costs, or estimated using a number of standard economic methods to account for the economic value of a project to the public. As with HEA and REA methods, monetization approaches are used to estimate offsets over a restoration project's lifespan.[17] For this ERP/EA, the Trustees used a monetizing approach to estimate

---

[16] As stated in Chapter 1, examples of ecological services include biological diversity, nutrient cycling, food production for other species, habitat provision, and other services that natural resources provide for each other.
[17] Monetization can incorporate a range of approaches and techniques that include directly attempting to estimate the consumer surplus associated with implementing the project, or the cumulative willingness-to-pay of a population for a project. Other more indirect approaches, for example benefits transfer, attempt to value the project using available information from other similar projects while making appropriate adjustments for differences in the project that have

Offsets for the Florida Boat Ramp Enhancement and Construction Project, described in Section 3.2.6.1.

The methods used to estimate Offsets for early restoration projects were implemented pursuant to the Framework Agreement. Offsets were negotiated with BP and reasonably reflect the estimated benefits for each project. Neither the amount of the Offsets nor the methods of estimation are precedent for assessing the gains provided by any other projects either during the early restoration process or in the assessment of total injury. In the context of early restoration under the Framework Agreement, the Trustees used best information and methodologies available in judging the adequacy of proposed restoration in satisfying OPA's mandates (see 15 C.F.R. Section § 990.25) while determining that agreements reached under the Framework Agreement are fair, reasonable, and in the public interest.

3.2.2 Louisiana Projects

For more than 10 years, Louisiana has used its RRP Program to solicit and integrate public input regarding the types of restoration projects that could best compensate the public for natural resource damages caused by oil spills. Following the Spill, Louisiana trustees engaged coastal stakeholders through a variety of public outreach and coordination efforts to discuss NRDA, the restoration planning process, and potential restoration projects specifically related to the Spill. In addition to the meetings discussed in Chapter 1 of this document, Louisiana trustees frequently met with stakeholders, both individually and collectively, to convey information and solicit suggestions. For example, the Coastal Protection and Restoration Authority of Louisiana and the Governor's Oyster Advisory Committee hold monthly public meetings in which these issues were, and continue to be, discussed.

From these recent outreach efforts, and the State's existing RRP Program, Louisiana compiled a list of potential projects for restoration of State natural resources injured as a result of the Spill. Project ideas received through June 25, 2011, were considered for the initial round of early restoration; however, the Louisiana trustees continue to accept restoration project ideas. To submit a project idea online, or to view the current list of project candidates, please visit http://losco-dwh.com. Projects submitted after June 25, 2011, as well as those projects not proposed for this initial phase of early restoration planning, may be considered for future stages of both early and comprehensive NRDA restoration planning.

Based on analysis of the selection criteria set forth in OPA NRDA regulations, the Framework Agreement and additional RRP Program-specific criteria[18], Louisiana proposed initial funding through the Framework Agreement for (1) the Lake Hermitage Marsh Creation – NRDA Early Restoration Project and (2) the Louisiana Oyster Cultch Project. These projects are consistent

---

already been valued and the project of interest for factors such as: project location, project scale, and characteristics of the affected populations.

[18] The additional Louisiana RRP Program criteria are:
   a.   Ability to Implement Project with Minimal Delay;
   b.   Degree to Which Project Supports Existing Strategies/Plans;
   c.   Project Urgency; and
   d.   Other Factors as Appropriate
(RRP Program FPEIS, NOAA et al. 2007b, p. 104).

with the Louisiana Coastal Master Plan, meet criteria outlined in OPA NRDA regulations, the Framework Agreement, and the RRP Program, and are consistent with the goal of compensating the public for natural resource injuries resulting from the Spill.

3.2.2.1   Lake Hermitage Marsh Creation – NRDA Early Restoration Project

3.2.2.1.1 Background and Project Description

The Lake Hermitage Marsh Creation – NRDA Early Restoration Project involves the creation of marsh within a project footprint known as the "Lake Hermitage Marsh Creation Project" developed for and funded through the Coastal Wetlands Planning, Protection and Restoration Act (CWPPRA) Program. This project substitutes approximately 104 acres of created brackish marsh for approximately 5-6 acres of earthen terraces that would otherwise have been constructed within the CWPPRA project boundary.

CWPPRA provides over $80 million per year for planning, design and construction of coastal restoration projects in Louisiana. Each year, a list of projects is selected for implementation, and funds are approved for engineering and design. The Lake Hermitage Marsh Creation Project (BA-42) was funded in 2006 as part of CWPPRA Priority Project List #15.

The Lake Hermitage Marsh Creation – NRDA Early Restoration Project is located within the Barataria Hydrologic Basin in Plaquemines Parish, Louisiana, to the west of the community of Pointe a la Hache, and northwest of the community of Magnolia (Figure 5). This basin was identified as a priority area for coastal restoration, and has been the focus of extensive study and project design and implementation.

The primary goals of the Lake Hermitage Marsh Creation base CWPPRA Project are (1) to restore the eastern Lake Hermitage shoreline to reduce erosion and prevent breaching into the interior marsh and (2) to re-create marsh in the open water areas south and southeast of Lake Hermitage. Specific objectives of the CWPPRA project are to: (1) create 549 acres of marsh by filling open-water areas and fragmented marsh with dredged material; (2) restore approximately 6,106 linear feet of the eastern Lake Hermitage shoreline; and (3) create 5 acres of emergent habitat by constructing 7,300 linear feet of earthen terraces. The terrace field proposed in the CWPPRA project consists of approximately 104 acres.

Figure 5. Lake Hermitage Marsh Creation – NRDA Early Restoration Project
location within the Barataria Basin.

Throughout the engineering and design phases of the CWPPRA project, the project team
considered incorporating an additional 104 acres of marsh creation in the footprint of the terrace
field. However, due to funding constraints, the project team completed final design of the
CWPPRA project with the 7,300 linear feet of earthen terraces (Figure 6).

The Lake Hermitage Marsh Creation – NRDA Early Restoration Project is designed to create
that additional 104 acres of brackish marsh in lieu of the earthen terraces included in the final
design of the base CWPPRA project (Figure 7). Marsh areas would be constructed entirely
within the base project's terrace boundary. Sediment would be hydraulically dredged from a
borrow area in the Mississippi River, and pumped via pipeline to create new marsh in the project
area. Over time, natural dewatering and compaction of dredged sediments should result in
elevations within the intertidal range which would be conducive to the establishment of emergent
marsh. The 104-acre fill area would be planted with native marsh vegetation to accelerate
benefits to be realized from this project. The estimated cost to implement the Lake Hermitage
Marsh Creation – NRDA Early Restoration Project is $14,400,000.

3.2.2.1.2 Selection Criteria

The Lake Hermitage Marsh Creation – NRDA Early Restoration Project would create new brackish marsh. The ecological services gained by this project are anticipated to help compensate for brackish marsh injuries or losses due to the Spill. The created marsh would be constructed in the Barataria Hydrologic Basin, which was heavily impacted by the Spill. Thus, this project has a clear nexus to resources injured by the Spill. *See* 15 CFR § 990.54 (a)(2); and 6(a)-(c) of the Framework Agreement. The project is technically feasible and utilizes proven techniques with established methods and documented results. Local, state and federal agencies have successfully implemented similar marsh creation projects in this region. For these reasons, the project has a high likelihood of success. See 15 CFR § 990.54 (a)(3); and 6(e) of the Framework Agreement.

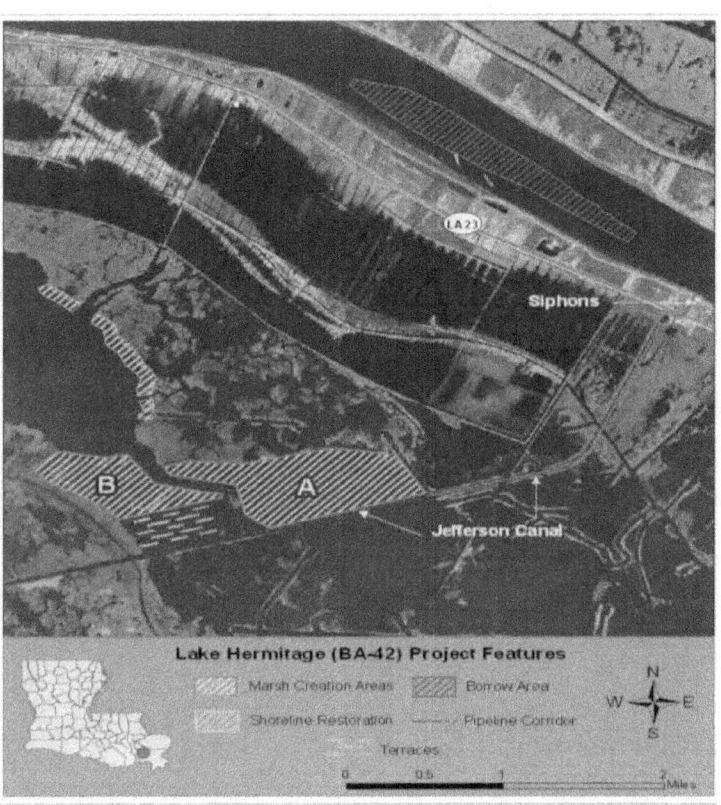

Figure 6. Lake Hermitage Marsh Creation CWPPRA Project (showing terrace field).

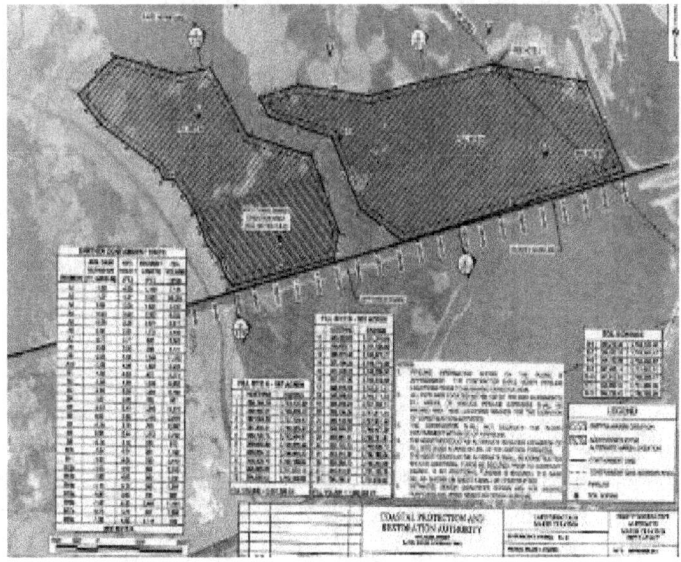

Figure 7. Lake Hermitage Marsh Creation – NRDA Early Restoration Increment.

The Lake Hermitage Marsh Creation – NRDA Early Restoration Project would be conducted at a reasonable cost. See 15 CFR § 990.54 (a)(1). The project is included as an alternative design in a CWPPRA project that is scheduled for completion within the year. As such, there exists a narrow window of opportunity in which the project can be constructed in conjunction with the construction of the CWPPRA project. See RRP Program FPEIS, NOAA et al., 2007b, p. 104. Constructing the project in conjunction with the construction of the CWPPRA project offers significant time and cost savings by achieving administrative and construction efficiencies. See RRP

31

Program FPEIS (NOAA et al., 2007b, p. 104); 15 C.F.R. § 999.54(b); and 6(e) of the Framework Agreement.

The project enhances a pre-existing restoration initiative and is consistent with broader restoration goals for Louisiana coastal wetlands. See RRP Program FPEIS (NOAA et al., 2007b, p. 104). The project is also consistent with anticipated long-term restoration needs and final restoration plans stemming from the Spill. See 6(d) of the Framework Agreement.

3.2.2.1.3   Performance Criteria, Monitoring and Maintenance

Project performance would be assessed by comparing quantitative monitoring results to predetermined performance standards that define the minimum physical or structural conditions deemed to represent normal and acceptable growth and development. The Trustees expect to conduct quantitative vegetation monitoring using ground surveys and also periodically conduct remote sensing of vegetation to obtain aerial coverage. The Trustees will also conduct annual inspections of the project to identify issues that may need correction. The monitoring program for this project would use quantitative standards for parameters such as percent live desirable vegetation to determine whether the project goals and objectives have been achieved, or whether corrective actions are required to meet the goals and objectives. Further details concerning the performance measures and monitoring would be developed prior to implementation of the project.

3.2.2.1.4   Offsets

For the purposes of negotiations of Offsets with BP in accordance with the Framework Agreement, the Trustees used Habitat Equivalency Analysis to estimate Offsets provided by the Lake Hermitage Marsh Creation – NRDA Early Restoration Project. Offsets reflect units of discounted service acre years (DSAYs) of emergent brackish salt marsh, and would be applied against emergent brackish salt marsh habitat injured by the Spill in the Barataria Hydrologic Basin as determined by the Trustees' total assessment of injury. In estimating DSAYs, the Trustees considered a number of factors, including, but not limited to, the time period that it would take for created marsh to provide different levels of ecological benefits, the time period over which the project would continue to provide benefits, and the ecological benefits of created marsh relative to existing marsh habitats that were not affected by the Spill. Total estimated Offsets for the Lake Hermitage Marsh Creation – NRDA Early Restoration Project are 518 DSAYs. In addition, the Trustees determined that approximately 25% of the Offsets (134 DSAYs) would be associated with highly productive marsh edge habitat, which is habitat along the land/water interface. These Offsets are reasonable for this resource and this project.

3.2.2.2 Louisiana Oyster Cultch Project

3.2.2.2.1   Background and Project Description

The Louisiana Oyster Cultch Project involves (1) the placement of oyster cultch onto approximately 850 acres of public oyster seed grounds throughout coastal Louisiana and (2) construction of an oyster hatchery facility that would serve to improve existing oyster hatchery operations and produce supplemental larvae and seed.

First, the Louisiana Department of Wildlife and Fisheries (LDWF) would contract for the placement of cultch material onto approximately 850 acres of public oyster seed grounds throughout coastal Louisiana, including 3-Mile Bay, Drum Bay, Lake Fortuna, South Black Bay, Hackberry Bay and Sister Lake (Figure 8). Cultch material consists of limestone rock,

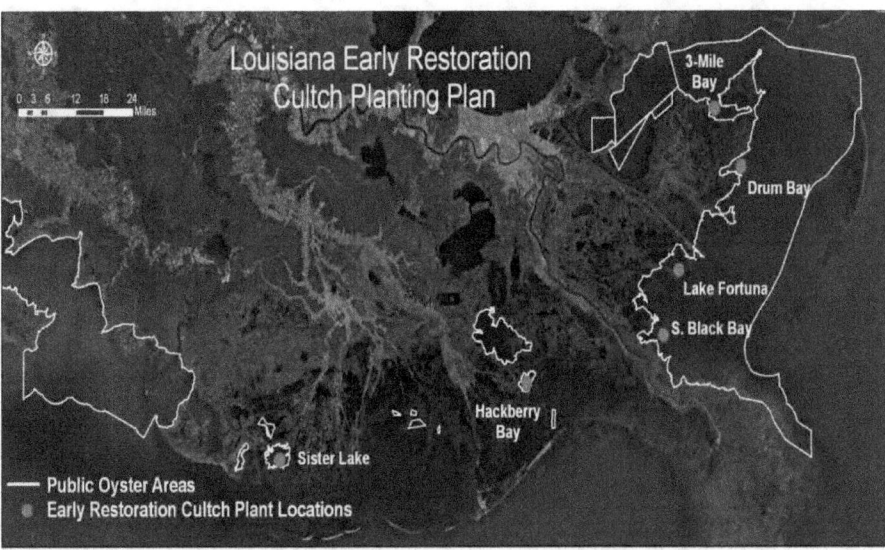

Figure 8. Louisiana oyster cultch planting locations.

crushed concrete, oyster shell and other similar material that, when placed in oyster spawning areas, provides a substrate on which free swimming oyster larvae can attach and grow into oysters. The cultch materials are planned to be placed at a planting density of 200 cubic yards/acre, although adjustments to this planting density may be made depending upon water bottom characteristics at the time of project implementation. The Louisiana Oyster Cultch Project would employ cultch planting approaches utilized by LDWF since 1917.

The second portion of the project involves constructing an oyster hatchery facility that would serve to improve existing oyster hatchery operations to help facilitate and expedite success of the cultch placement. Since the Spill, spat fall in some of the areas impacted by the Spill has been lower than average. In order to provide a supplemental source of oyster larvae and oyster seed, LDWF, in partnership with Louisiana Sea Grant, would contract to construct a new building adjacent to the existing Sea Grant oyster hatchery located at the LDWF facility on Grand Isle, Louisiana. Larvae produced at the hatchery can be released into the water directly over cultch material or be remotely set on oyster cultch to create oyster seed. The new facility would be located next to the LDWF Research Lab at a site leased by Louisiana State University, located at 133 Port Drive in Grand Isle, Louisiana. The site, which is currently undeveloped, is approximately 20,186 square feet, and is owned by the Grand Isle Port Commission and leased by Louisiana State University. Louisiana State University plans to construct an additional building at this site prior to construction of the hatchery facility; this building is not part of this Early Restoration Project. The footprint of the hatchery building is proposed to be approximately 8400 sq ft. Parking will be available onsite. Oyster hatchery activities currently housed at the LDWF Research Lab in Grand Isle, Louisiana will relocate to the new hatchery building once it is constructed.

Hatchery operations would include broodstock maintenance, algal cultivation, larvae production, and a nursery system. Broodstock (adult oysters used in oyster breeding), which would continue to be located at the LDWF Research Lab, are collected in Louisiana waters. Broodstock are critical to hatchery operations as they ensure a source of males and females of specific genetic traits that are used to produce larvae and eventually seed. Algae are the primary source of food for both larvae and adult oysters. At the hatchery, broodstock would be thermally induced to spawn. The resulting fertilized eggs would be added to a tank and allowed to hatch. The free swimming larvae move up and down the water column feeding on algae, and grow and develop (after approximately 15 days) into "eyed" larvae that look like a small clam. Once an oyster reaches the eyed larvae stage it is ready to settle or spat onto hard substrate. Once oyster larvae reach the proper age and size they can be broadcast onto suitable coastal areas (i.e., cultched areas), or encouraged to settle (set) onto small pieces of shell in the hatchery. After the larvae set on the shell, they are called "spat." Spat can be grown into seed in the hatchery nursery system. The nursery system consists of a series of upweller silos, which are columns (2' x 1.5') through which water is pushed from the bottom. The system would use the water from, and would replace the water to, the bay immediately adjacent to the new facility. Planned capacity for the hatchery system is approximately 8,000 gallons of water per day from April through October. When oysters reach approximately 1 inch in length they would be moved to a suitable growout area (i.e., public seed grounds). The facility is designed to produce 1 billion eyed larvae per season.

The estimated cost to implement the Louisiana Oyster Cultch Project is $15,582,600.

3.2.2.2.2   Selection Criteria

The goal of the Louisiana Oyster Cultch Project is to produce seed-sized and sack-sized oysters on public oyster seed grounds. Oysters were exposed to oil, dispersant, as well as response activities undertaken to prevent, minimize, or remediate oiling from the Spill. Thus, the nexus to resources injured by the Spill is clear. See 15 C.F.R. § 990.54 (a)(2). See also 6(a)-(c) of the Framework Agreement. The project employs cultch planting methods and techniques that the State of Louisiana has used for decades to manage its oyster resource. Therefore, the project is both technically feasible and carries a high probability of success. See 15 C.F.R. § 990.54 (a)(3); and 6(e) of the Framework Agreement.

The Louisiana Oyster Cultch Project can be conducted at a reasonable cost and may be implemented by the State with minimal delay. See 15 C.F.R. § 990.54 (a)(1); RRP Program FPEIS (NOAA et al., 2007b, p. 104); and 6(e) of the Framework Agreement. The project supports existing restoration initiatives and strategies and is consistent with anticipated long-term restoration needs and anticipated final restoration plans stemming from the Spill. See RRP Program FPEIS (NOAA et al. 2007b, p. 104); and 6(d) of the Framework Agreement.

3.2.2.2.3   Performance Criteria, Monitoring and Maintenance

Project performance would be assessed through physical and biological monitoring of oyster cultch plants. The monitoring program would determine whether the project goals and objectives have been achieved, or whether corrective actions are required to meet the goals and objectives.

Biological monitoring parameters would consist of oyster metrics including density, size, and spat settlement in cultch plants. This monitoring would be consistent with the oyster monitoring protocols used by the LDWF in their annual oyster stock assessment activities. Oyster cultch plant maintenance would consist of periodic evaluation of cultch coverage within the placement boundaries and could include cultch replenishment, if feasible. Cultch material is expected to be lost over time due to weather events, relay of seed-sized oysters, harvest activity, etc., and the Trustees' calculations of benefits have taken into account this expected loss over time.

3.2.2.2.4 Offsets

For the purposes of negotiations of Offsets with BP in accordance with the Framework Agreement, the Trustees used Resource Equivalency Analysis to estimate Offsets for the Louisiana Oyster Cultch Project, resulting in expected production of oysters on cultch material over time. Offsets reflect estimated kilograms of oysters produced, and would be applied against oyster injuries in coastal Louisiana injured by the Spill as determined by the Trustees' total assessment of injury. The Trustees considered a number of factors in estimating oyster production, including, but not limited to, typical oyster production in the project area, estimated project life span and size of the project. Total estimated Offsets for the Louisiana Oyster Cultch Project are 4,000,000 discounted kilogram-years (Dkg-Y) of oyster secondary production.[19] These Offsets are reasonable for this resource and this project.

3.2.3 Mississippi Projects

3.2.3.1 Mississippi Oyster Cultch Restoration

3.2.3.1.1 Background and Project Description

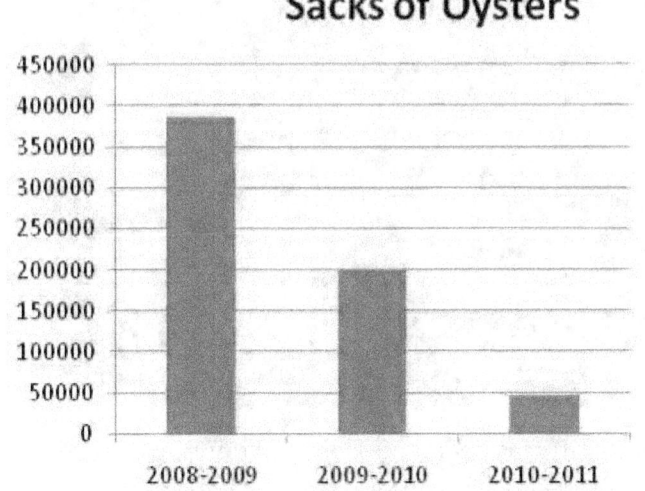

Figure 9. Oyster production (in Sacks of Oysters harvested), 2008 to 2011. Source: MDMR, 2010; MDMR, 2011a, 2011b.

For over a hundred years shell/cultch plants in Mississippi have proven to be successful in growing new and refurbishing damaged oyster cultch areas. The state of Mississippi has approximately 12,000 acres of total cultch areas, including about 9,000 acres of oyster cultch area which can be harvested in the Mississippi Sound, and about 3,000 acres of cultch areas closed to harvest. Once clean oyster cultch has been planted and larval oysters become attached, oysters may grow to legally harvestable size in 18 to 36 months. Mississippi typically does not open oyster areas to harvest until five or six years after cultch placement. Figure 9 depicts oyster production in the Mississippi sound from 2008 to 2011.

---

[19] Ash-Free-Dry-Weight of oyster tissue. These Offsets are applicable first to any oyster injuries in Louisiana and if any surplus remains, to nearshore benthic invertebrate injuries in Louisiana.

The goal of this project is to restore and enhance oyster cultch areas in the marine waters of the Mississippi Sound in Hancock and Harrison counties. Oyster cultch plant areas are routinely surveyed to identify potential enhancement and restoration opportunities. This project would restore and enhance approximately 1,430 acres of the oyster cultch areas within the Mississippi Sound in Hancock and Harrison counties (Figure 10). Cultch material (oyster shell, limestone or crushed concrete, or some combination thereof) would typically be deployed at a rate of 100 cubic yards per acre within existing oyster cultch area footprints with adjustments for site conditions as needed. Cultch deployment sites will be screened prior to cultch placement. Locations that are not safe or suitable for oyster production would not be used. Deployment would occur in Fall 2012, Spring 2013 and Fall 2013.

The estimated cost for this project is $11,000,000.

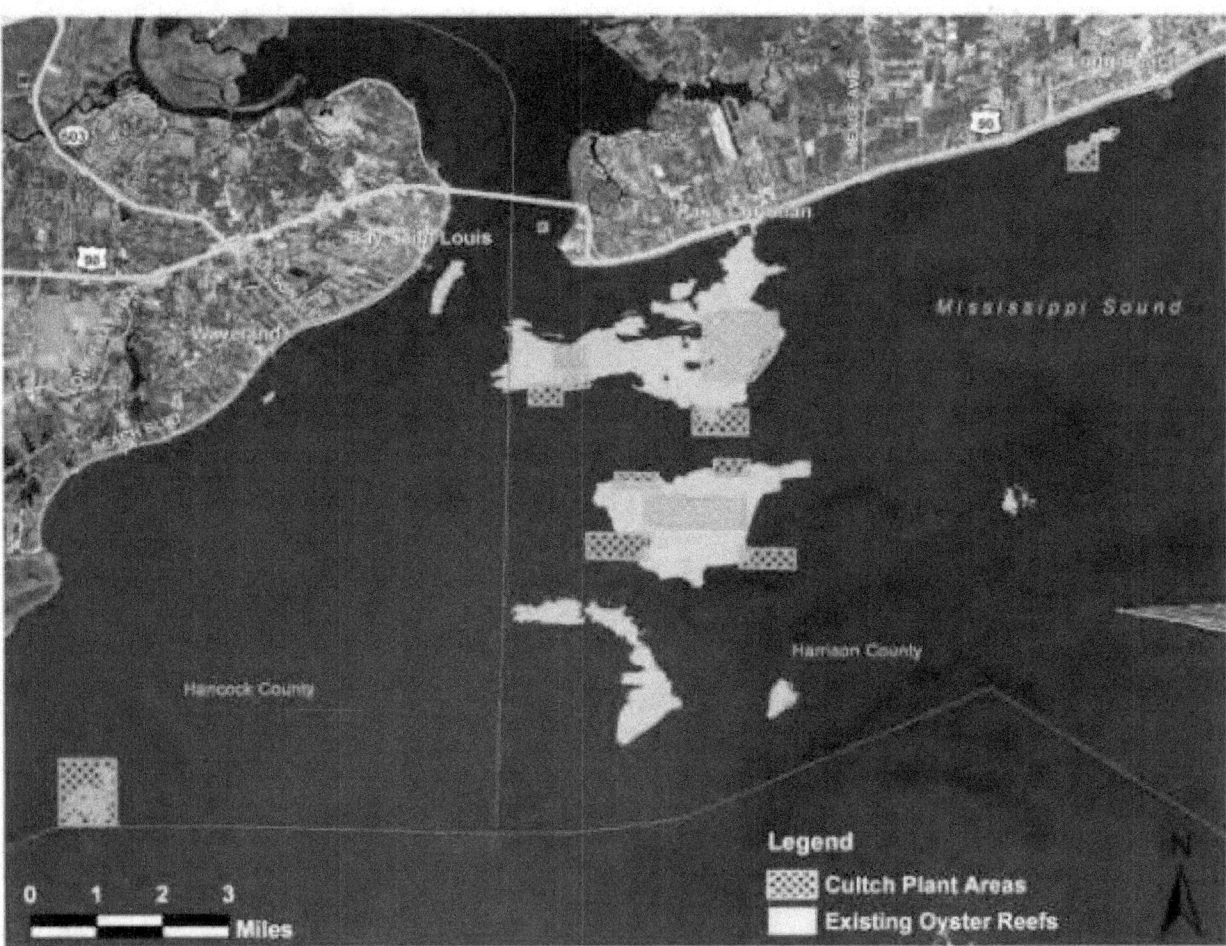

Figure 10. Mississippi Sound oyster growing areas in Hancock and Harrison counties.

3.2.3.1.2 Selection Criteria

Oyster reef restoration was suggested as a restoration measure during NOAA's public scoping meetings for the Deepwater Horizon PEIS, and also submitted as restoration project(s) on the

NOAA website (http://www.gulfspillrestoration.noaa.gov). The Mississippi Oyster Cultch Restoration project would restore injured oyster reefs and/or compensate for interim losses of such natural resources within the Mississippi Sound for impacts to oysters exposed to oil, dispersant, and/or response activities undertaken to prevent, minimize, or remediate oiling from the Spill. Thus, the nexus to resources injured by the Spill is clear. See 15 C.F.R. § 990.54 (a)(2). See also 6(a)-(c) of the Framework Agreement.

Restoration through typical oyster cultch placements start with natural spat settlement. The cultch restoration would result in an oyster reef within 3 to 5 years. The project would be implemented by the Trustee coordinating with the Mississippi Department of Marine Resources (MDMR), which has a long-standing oyster cultch restoration program. Additionally, monitoring and management of the oyster resources would ensure the likelihood of success of this and future oyster bed restoration in the Mississippi Sound. Therefore, the project is both technically feasible and carries a high probability of success. See 15 C.F.R. § 990.54 (a)(3); and 6(e) of the Framework Agreement. The project can be conducted at a reasonable cost and may be implemented by the Trustee with minimal delay. See 15 C.F.R. § 990.54 (a)(1); and 6(e) of the Framework Agreement. Accordingly, the Mississippi Oyster Cultch Restoration project meets the evaluation criteria for the Framework Agreement and OPA discussed in Section 1.6.

3.2.3.1.3   Performance Criteria, Monitoring and Maintenance

Project performance would be assessed through physical and biological monitoring of oyster cultch plants. The monitoring program would determine whether the project goals and objectives have been achieved, or whether corrective actions are required to meet the goals and objectives. Biological monitoring will include typical oyster metrics (i.e., density, size, and spat settlement).

Oyster cultch plant maintenance would consist of remote sensing of cultch coverage within the placement boundaries and cultch replenishment, as necessary. Cultch material may be lost over time due to weather events, harvest activity, etc. Mid-course enhancements would include additional cultch placement in areas of cultch loss.

3.2.3.1.4   Offsets

For the purposes of negotiations of Offsets with BP in accordance with the Framework Agreement, the Trustees used Resource Equivalency Analysis to estimate Offsets for Mississippi Oyster Cultch Restoration, resulting in expected production of oysters on cultch material over time. Offsets reflect estimated kilograms of oysters produced, and would be applied against oyster injuries in Mississippi Sound injured by the Spill as determined by the Trustees' total assessment of injury. The Trustees considered a number of factors in estimating oyster production, including, but not limited to, typical oyster production in the project area, estimated project life span and size of the project. Total estimated Offsets for Mississippi Oyster Cultch Restoration is 2.0 million Discounted Kilogram (Dkg) Years of oyster biomass.[20] These Offsets are reasonable for this resource and this project.

---

[20] Ash-Free-Dry-Weight of oyster tissue. These Offsets are applicable first to any oyster injuries in Mississippi and if any surplus remains, to nearshore benthic invertebrate injuries in Mississippi.

3.2.3.2   Mississippi Artificial Reef Habitat

3.2.3.2.1   Background and Project Description

The Mississippi Artificial Reef Habitat project proposes to deploy nearshore artificial reefs in the Mississippi Sound. Nearshore artificial reefs provide valuable hardbottom habitat with foraging and shelter sites for various species of larvae and sessile epifauna and infauna. Currently there are 67 existing nearshore artificial reef areas that are each approximately 3 acres in size. At present, approximately half of these existing reef areas have a low profile and consist of crushed concrete or limestone. The locations of Mississippi's existing nearshore artificial reefs are shown in Figure 11. With the Mississippi Artificial Reef Habitat project, approximately 100 acres of crushed limestone would be added to the 201-acre footprint of the existing reef areas or hard substrate habitats. The resulting artificial reefs would consist of low profile reefs 4 to 6 inches above the seafloor.

The estimated cost for this project is $2,600,000.

Figure 11. Mississippi's existing nearshore artificial reefs.

3.2.3.2.2 Selection Criteria

Artificial reefs were suggested as restoration measures during NOAA's public scoping meetings for the Deepwater Horizon PEIS, and also submitted as restoration project(s) on the NOAA website (http://www.gulfspillrestoration.noaa.gov). The Mississippi Artificial Reef Habitat project would restore injured secondary productivity in the Mississippi Sound, resulting from exposure to oil, dispersant, and/or response activities undertaken to prevent, minimize, or remediate oiling from the Spill. Thus, the nexus to resources injured by the Spill is clear. See 15 C.F.R. § 990.54 (a)(2). See also 6(a)-(c) of the Framework Agreement.

The project would be implemented by the Trustee in coordination with MDMR, which has a long-standing artificial reef program which includes placement, management, and monitoring of reef areas. Artificial reef material placement sites will be screened prior to deployment. Deployment will be limited to areas that are suitable and safe. All effort would be made to avoid existing environmentally sensitive areas including any existing benthic communities. Therefore, the project is both technically feasible and carries a high probability of success. See 15 C.F.R. § 990.54 (a)(3); and 6(e) of the Framework Agreement. The project can be conducted at a reasonable cost and may be implemented by the State with minimal delay. See 15 C.F.R. § 990.54 (a)(1); and 6(e) of the Framework Agreement.

3.2.3.2.3 Performance Criteria, Monitoring and Maintenance

The Mississippi Artificial Reef Habitat project involves the placement of a layer of crushed limestone only within the existing nearshore reef site footprints in Mississippi. Project performance would be measured through a physical and biological monitoring program. The Trustee, in coordination with the University of Southern Mississippi Gulf Coast Research Laboratory (USM GCRL), would conduct biological monitoring of the nearshore reefs. Project performance will be measured through a physical and biological monitoring program. Findings from the monitoring will be used to determine reef success, performance, expected benefits, and maintenance and management activities. Physical monitoring of the structure and integrity of nearshore reef systems will be based on observations during biological monitoring.

3.2.3.2.4 Offsets

For the purposes of negotiations of Offsets with BP in accordance with the Framework Agreement, the Trustees used Resource Equivalency Analysis to estimate Offsets for Mississippi Artificial Reef Habitat project, resulting in expected production of invertebrate infaunal and epifaunal biomass at nearshore artificial reefs. Offsets reflect estimated kilograms of biomass produced, and would be applied against secondary productivity injuries in Mississippi Sound from the Spill as determined by the Trustees' total assessment of injury. The Trustees considered a number of factors in estimating biomass production, including, but not limited to, typical productivity in the project area, estimated project life span and size of the project. Total estimated Offsets for the Mississippi Artificial Reef Habitat project are 763,609 Dkg-Ys of invertebrate infaunal and epifaunal biomass[21] at nearshore artificial reefs in Mississippi. These Offsets are reasonable for this resource and this project.

3.2.4 Alabama Project

3.2.4.1 Marsh Island (Portersville Bay) Marsh Creation

The Marsh Island (Portersville Bay) Restoration Project involves the creation of salt marsh along Marsh Island, a state-owned island in the Portersville Bay portion of Mississippi Sound, Alabama. This project would add 50 acres of salt marsh to the existing 24 acres of Marsh Island through the construction of a permeable segmented breakwater, the placement of sediments and the planting of native marsh vegetation. Additionally, this project would protect the existing salt marshes of Marsh Island, which have been experiencing significant losses due to chronic erosion. Without the breakwater, the existing marsh would be completely washed away in approximately 15 years.

3.2.4.2 Background and Project Description

The Marsh Island (Portersville Bay) Marsh Creation Project is located within the Portersville Bay portion of Mississippi Sound in south Mobile County, Alabama (Figure 12). This area was identified as a top priority for coastal restoration by Alabama and its natural resource partners, and has been the focus of a number of recent restoration projects. The Marsh Island (Portersville Bay) Marsh Creation Project area specifically has experienced tremendous loss of emergent wetlands. An analysis of NOAA shoreline vectors and historic aerial imagery conducted by the Alabama Department of Conservation and Natural Resources indicates that Marsh Island has decreased in size by approximately 50% since 1958 and has a current shoreline recession rate of 5-10' per year (Figure 13).

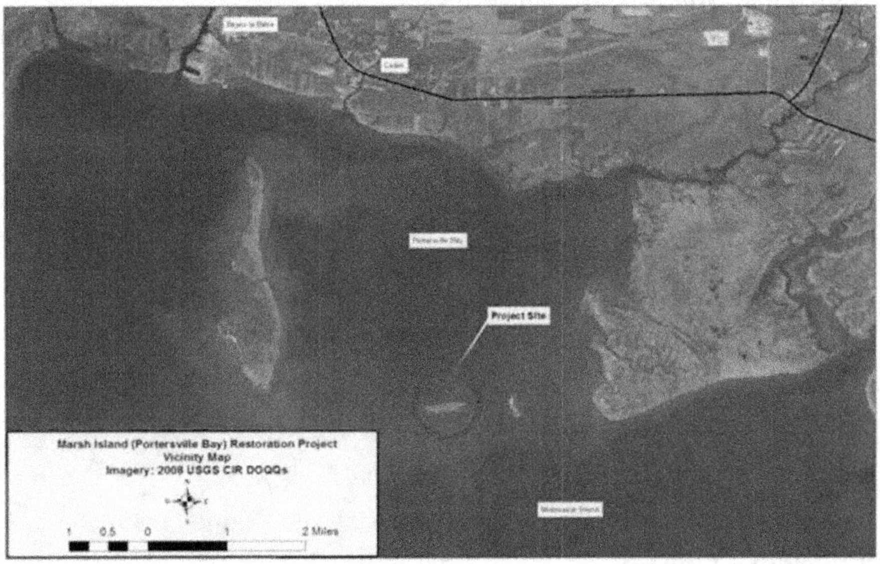

Figure 12. Marsh Island (Portersville Bay) Restoration Project, Portersville Bay, Alabama.

The primary goals of the Marsh Island (Portersville Bay) Marsh Creation Project are (1) to protect the southern shoreline of the island to reduce and/or prevent further erosion of the

---

[21] Ash-Free-Dry-Weight of Secondary Production of invertebrate infauna and epifaunal biomass.

existing salt marsh and (2) to re-create salt marsh in the open water areas north of the remainder of the island.

To implement these goals, the project would: (1) install approximately 5,700 linear feet of permeable segmented breakwater; (2) place approximately 245,000 cubic yards of dredged materials to create 50 acres of marsh by filling open-water areas with dredged material; and (3) plant approximately 312,500 native vegetation plugs (see Figure 14). Additionally, through the natural dewatering and compaction of dredged sediments and the use of a marsh buggy, approximately 5,000 linear feet of tidal creeks would be created, connecting existing tidal creeks to the newly created marsh and to Mississippi Sound.

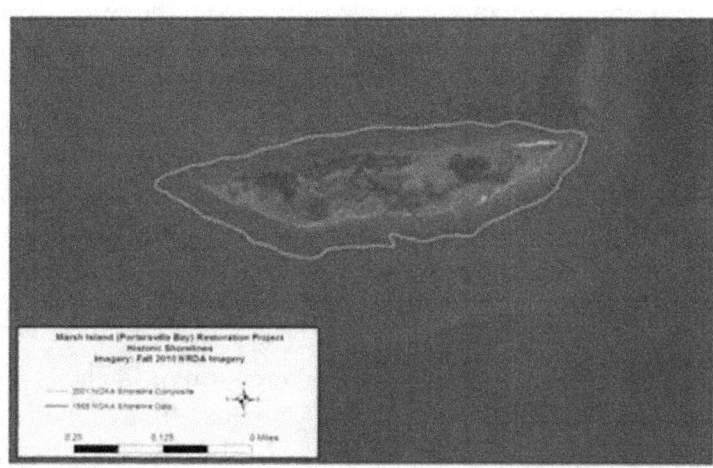

Figure 13. Marsh Island erosion, 1958-present.

The estimated cost for this project is $11,280,000.

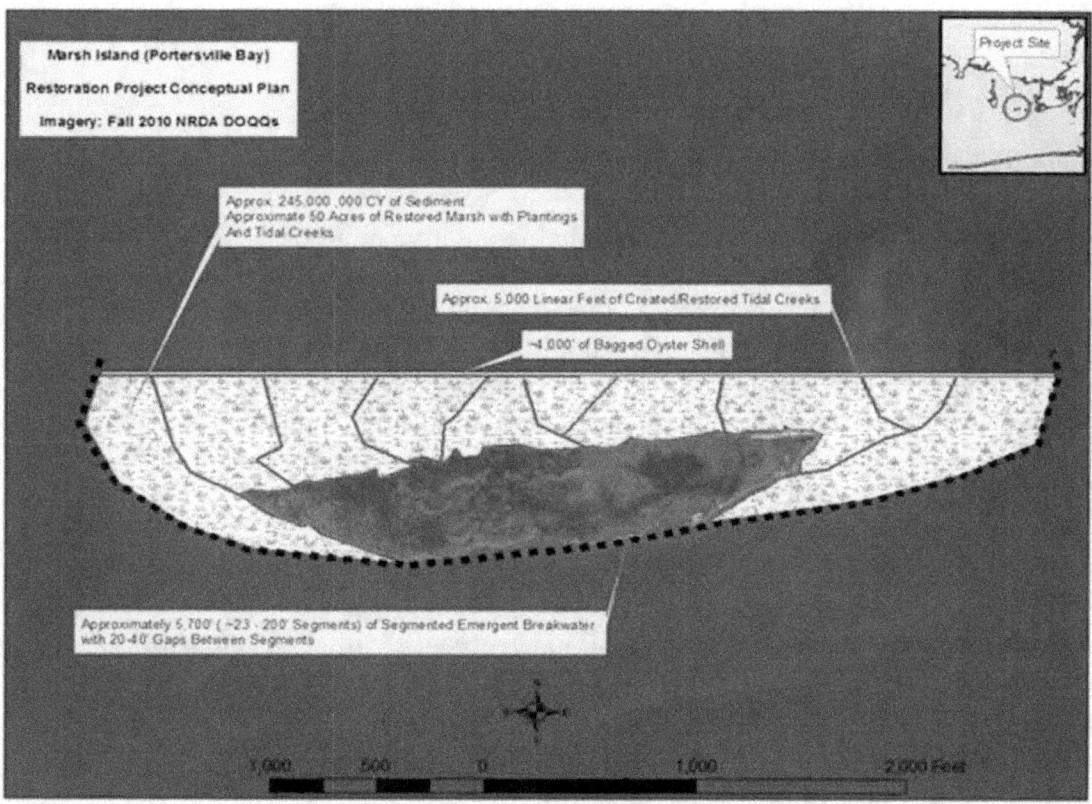

Figure 14. Marsh Island (Portersville Bay) Restoration Project conceptual design.

3.2.4.3 Selection Criteria

Marshes in Alabama were exposed to oil, dispersant, and response activities undertaken to prevent, minimize, or remediate oiling from the Spill. As such, a marsh restoration project is one of Alabama's priorities for early restoration. The goal of the Marsh Island (Portersville Bay) Marsh Creation Project is to create a structurally robust, emergent salt marsh designed to provide maximum salt marsh ecological benefits as soon as practicable. Ecological services gained by the created marsh would help compensate for salt marsh injuries or losses due to the Spill. Marshes in Mississippi Sound were impacted by the Spill even though oil did not come ashore on Marsh Island itself. This type of project has been completed in Alabama in the past and the Trustees felt comfortable that implementing such a project would help restore or replace marsh services like those lost.

A number of marsh restoration and creation projects have been submitted to the Trustees for consideration. These projects for Alabama all have merit and would have the potential to address resource injuries associated with the Spill. However, based on the criteria identified in OPA regulations (15 C.F.R. § 990.54), which are also consistent with the guidance provided in the Framework Agreement, the Trustees determined that the Marsh Island project could serve as one of the best projects to propose for Phase I early restoration. This restoration project would provide for the protection of the existing marsh and creation of new marsh, thereby providing ecological service gains to help compensate for injuries to or losses of salt marsh in Alabama caused by the Spill. This project is similar to other restoration projects that have occurred in coastal Alabama and the likelihood of success is high. It is also cost-effective and has a lengthy projected lifespan. The Trustees do not anticipate any adverse impacts associated with this project and there is no significant risk to human health and safety.

3.2.4.4 Performance Criteria, Monitoring and Maintenance

Project performance would be assessed by comparing quantitative monitoring results to predetermined performance standards that define the minimum physical or structural conditions deemed to represent normal and acceptable growth and development (e.g., elevation and colonization of native emergent vegetation). The monitoring program for this project would use these standards to determine whether the project goals and objectives have been achieved, or whether corrective actions are required to meet the goals and objectives. Details concerning the performance measures and monitoring would be developed prior to implementation of the project.

3.2.4.5 Offsets

For the purposes of negotiations of Offsets with BP in accordance with the Framework Agreement, the Trustees used Habitat Equivalency Analysis to estimate Offsets provided by the Marsh Island Project. Offsets reflect units of discounted service acre years (DSAYs) of salt marsh, and would be applied against salt marsh habitat along the coast of Alabama injured by the Spill as determined by the Trustees' total assessment of injury. In estimating DSAYs, the Trustees considered a number of factors, including, but not limited to, anticipated protection of Marsh Island's existing acres of marsh provided by the project, new marsh created by the project, the time period that it would take for created marsh to provide different levels of ecological benefits, the time period over which the project would continue to provide benefits, and the ecological benefits of created marsh relative to existing marsh habitats that were not affected by the Spill. Total estimated Offsets for the Marsh Island Project are 540 DSAYs. These Offsets are reasonable for this resource and this project.

3.2.5 DOI Project

3.2.5.1 Alabama Dune Restoration Cooperative Project

The City of Gulf Shores, City of Orange Beach, Gulf State Park, Bon Secour NWR and the BLM form the largest group of coastal land owners along the Alabama Gulf Coast. These owners collectively own and/or manage more than 20 miles of dune habitat. The Alabama Dune Restoration Cooperative Project would result in the formation of a partnership, the Coastal Alabama Dune Restoration Cooperative (CADRC), to restore dune habitat injured by the Spill. The CADRC would restore approximately 55 acres of primary dune habitat in Alabama by planting native dune vegetation and installing sand fencing. The project would help prevent erosion by restoring a "living shoreline": a coastline protected by plants and associated dunes rather than hard structures. These natural resources provide habitat to wildlife and increase the storm protection to both habitat and human resources.

3.2.5.2 Background and Project Description

Approximately 680,000 native plants would be planted within designated project areas (Figure 15). Proportions of plants would include approximately 70% sea oats grasses, 20% panic and

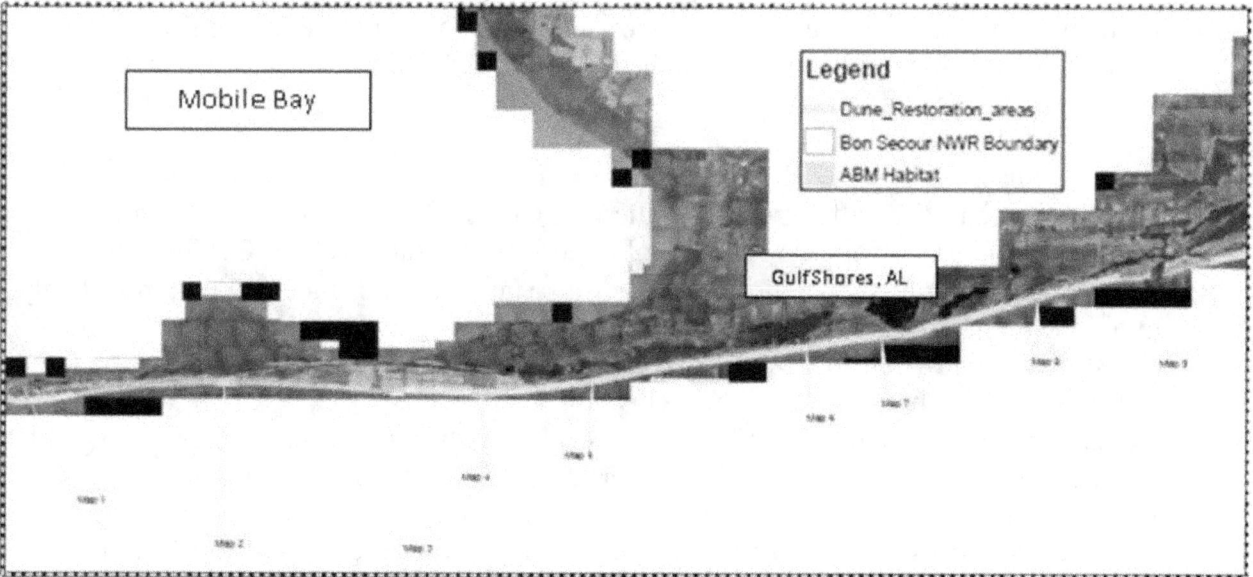

Figure 15. Alabama Dune Restoration Cooperative Project planting/fencing areas.

smooth cord grasses, and 10% ground cover plants (sea purslane, beach elder, white morning glories and railroad vine) to maximize sand stabilization and limit wind erosion. All plants would be grown from seeds or cuttings from the Alabama or North Florida coast to ensure appropriate genetic stocks are used in the project. Plants would be installed at 18-inch centers and 6 inches deep to ensure that sufficient moisture is available to roots. Planting would be targeted for the March-June time frame. Slow release fertilizer would be added during plant installation and plants would be periodically watered, as needed, to facilitate establishment.

Protective sand fencing would be installed around dunes on BLM property at the Our Road tract and in areas managed by the cities of Orange Beach and Gulf Shores. Sand fencing would be installed according to the approved Alabama Department of Environmental Management guidelines seaward of existing dunes, or as needed to promote sand accumulation in areas without established dunes.

No new access roads or staging areas would be built as part of this project. Vehicles would use existing roads and parking areas. All participants involved in the project would follow guidelines and designated access points established by DOI and its partners to minimize foot traffic and human presence across ecologically sensitive areas.

Informative dune restoration signage would be placed on the project area at a rate of 10-25 signs per mile in an effort to reduce human disturbance of restored areas.

All aspects of the project would be implemented using the best management practices described below.

Alabama beach mouse:
- To minimize potential impacts during installation of dune plants and sand fencing, all possible Alabama beach mouse burrows will be flagged under the supervision of a qualified biologist. These flagged burrows will be avoided during the project.
- If an Alabama beach mouse burrow cannot be avoided, the qualified biologist will stop installation activities and consult with the U.S. Fish and Wildlife Service Daphne Ecological Service Office.

Loggerhead sea turtle
- Restoration activities will be subject to the following mitigation measures that are designed to minimize impacts to nesting Loggerhead sea turtles (May-October).
- Restoration activities should ideally occur from March through June and will most likely avoid the highest loggerhead sea turtle nesting/hatching activity that occurs from mid-June through mid-August. However, when restoration occurs during nesting season the precautions described below will be followed.
- Actual installation of dune plants and sand fencing will occur during daylight hours and will therefore not impact nesting females or hatchlings that are active during the evening hours. Additionally, no restoration equipment will be left on the beach overnight. Likewise, all Loggerhead sea turtle nests in the project area are marked each morning by survey crews by 9 am. Therefore, restoration crews shall not begin work in an area until after it is cleared by the survey crews. If a nest occurs in a restoration area the nest will be avoided by no less than ten feet.

- To minimize potential impacts of the sand fencing on sea turtle nesting after installation, the Alabama Department of Natural Resource minimal distance guidelines for sand fence installation will be followed.

Kemp's Ridley sea turtle

Restoration activities will be subject to the following mitigation measures that are designed to minimize impacts to nesting Kemp's Ridley sea turtle nesting activities (May-October):

- Kemp's Ridley sea turtles infrequently nest in Alabama and often nest and hatch during daylight hours. Therefore, all restoration staff will be trained by a qualified biologist to avoid nesting and hatching Kemp's Ridley sea turtles by maintaining a minimum distance of 200 feet from the nesting or hatching Kemp's Ridley sea turtles. Additionally, the restoration crews will be required to immediately report the location of any nesting or hatching Kemp's Ridley sea turtles to a Bon Secour National Wildlife Refuge wildlife biologist, who will mark the nests. If there are no individuals (adults or hatchlings) present on the surface of the beach, then a marked nest will be avoided by no less than ten feet. Lastly, no restoration equipment will be left on the beach overnight.
- To minimize potential impacts of the sand fencing on sea turtle nesting after installation the Alabama Department of Natural Resource minimal distance guidelines for sand fence installation will be followed.

Piping plover

Restoration activities will be subject to the following mitigation measures that are designed to minimize impacts to piping plovers and associated overwintering habitat:

- Restoration activities should ideally occur from March through June and will most likely avoid piping plover overwintering in Alabama from September through April. However, when restoration occurs during the overwintering season the precautions described below will be followed.
- Vehicles used for restoration on the sandy beach south of the primary dune shall not exceed 10 mph.
- Heavily occupied habitat will be marked by qualified biologists and will be avoided by restoration staff until the piping plovers leave the area.

Snowy plover

Restoration activities will be subject to the following mitigation measures that are designed to minimize impacts to snowy plovers and associated nesting habitat:

- Each week a qualified biologist will survey the active restoration sites for snowy plover activity during nesting season. Areas of consistent activity will be flagged off and avoided by restoration crews until the birds leave the area.

The estimated cost for this project is $1,480,000.

3.2.5.3 Selection Criteria

Primary vegetated dune habitat located in the Bon Secour National Wildlife Refuge (NWR), Bureau of Land Management (BLM) Fort Morgan properties, and other parts of Alabama was injured by exposure to *Deepwater Horizon* oil and/or the extensive use of all-terrain vehicles, heavy equipment and personnel on beaches during response activities undertaken to prevent, minimize and/or remediate oiling. This habitat is located along seaward, frontal dunes, and characterized by a mixture of open sandy areas, grasses and forbs. The vegetative community is typically dominated by plants such as sea oats, panic grass, beach morning-glory, and seashore elder. The natural succession of dune vegetation and the seaward migration of the dune ecosystem were impeded for almost 2 years due to the necessity to provide access to the Alabama beaches during the Spill event. The Alabama Dune Restoration Cooperative Project will directly restore primary dune habitat injured by the Spill.

The Alabama Dune Restoration Cooperative Project meets the evaluation criteria for the Framework Agreement and OPA. The project would restore the equivalent of natural resources (vegetated dune habitat) injured by the Spill (*See* CFR § 990.54(a)(2) and Sections 6a-6c of the Early Restoration Framework Agreement) using established techniques. Trustees and their partners have successfully completed similar dune habitat restoration projects along the northern Gulf coast using these same protocols for decades. Cost estimates are based on similar past projects. As a result, the project is considered feasible, cost effective, and consistent with long-term restoration needs (*See* CFR § 990.54(a)(1),(3),(4) and Sections 6d-6e of the Early Restoration Framework Agreement). Over half of the dune restoration project is within Alabama beach mouse habitat and would assist in restoring a portion of the needs of the beach mouse, thus benefiting more than one natural resource and/or service. Monitoring and management of the restored habitat would enhance the likelihood of success of the project and the natural progression of the dunes.

3.2.5.4 Performance Criteria, Monitoring and Maintenance

Large storm events, severe drought and other activities could potentially negatively affect the success of plantings and sand fencing in dune habitat restoration. The CADRC would monitor plant and fence installations to evaluate project success. The plantings would be monitored for 90-days to assess plant survival. This project includes a provision for 90 day/80% survival guarantee and any plants lost during this time would be replaced. Following the initial performance monitoring, CADRC members would monitor the effectiveness of the plantings and sand fence installation by tracking changes in dune expansion or establishment. Large storm events and severe drought are the primary threats to project success.

3.2.5.5 Offsets

For the purposes of negotiations of Offsets with BP in accordance with the Framework Agreement, the Trustees used Habitat Equivalency Analysis to estimate Offsets provided by the Alabama Dune Restoration Cooperative Project. Offsets reflect units of discounted service acre years (DSAYs) of dune habitat, and would be applied against the Trustees' assessment of total injury to primary dune habitat along the Alabama coast injured by the Spill as determined by the Trustees' total assessment of injury. In estimating DSAYs, the Trustees considered a number of factors, including, but not limited to, benefits of revegetating primary dune habitat, the time period that it would take for revegetated habitat to provide different levels of ecological benefits, estimated project life span, potential impact of hurricanes and drought, and the ecological benefits of created dune relative to existing dune habitats that were not affected by the Spill.

Total estimated Offsets for the Alabama Dune Restoration Cooperative Project is 240 DSAYs. Because 55% of the restoration project area occurs in habitat utilized by the federally-endangered Alabama beach mouse (*Peromyscus polionotus ammobates*), 55% of the credits (132 DSAYs) can be used to offset injuries to primary vegetated dune habitat in Alabama utilized by the Alabama beach mouse. These Offsets are reasonable for this resource and this project.

3.2.6 Florida Projects

Following the Spill, Florida trustees engaged coastal governments, stakeholders, non-government organizations, state and regional agencies, and the public through a variety of public outreach and coordination efforts to discuss NRDA, the restoration planning process, and potential restoration projects specifically related to the Spill. In addition to the meetings discussed in Chapter 1 of this document, state Trustees frequently met with local municipalities and county governments, both individually and collectively, to convey information and solicit suggestions. Numerous conference calls were also held to coordinate with these government officials.

Based on outreach efforts Florida compiled a list of potential projects for restoration of natural resources and services injured, including human use services. Over 214 project ideas were received through September 21, 2011, and have been evaluated for the initial round of early restoration. The Florida Trustees will continue to accept restoration project ideas. To submit a project idea online, or to view the current list (List 1) of project candidates, please visit http://www.dep.state.fl.us/deepwaterhorizon/projects.htm. Projects not proposed for this initial phase of early restoration planning will be considered for future stages of both early and long-term restoration.

Based on analysis of the selection criteria set forth in OPA NRDA regulations, the Framework Agreement and additional Florida early restoration specific criteria, Florida is proposing the following initial early restoration projects: (1) the Florida Boat Ramp Enhancement and Construction Project and (2) the Florida (Pensacola Beach) Dune Restoration Project. These projects are consistent with the goal of restoring or replacing ecological and human use service losses resulting from the Spill.

3.2.6.1 Florida Boat Ramp Enhancement and Construction Project

The Florida Boat Ramp Enhancement and Construction Project will provide boaters enhanced access to public waterways within Pensacola Bay, Perdido Bay, and offshore areas. The project involves enhancement of public boat ramps in Escambia County, including repairs to existing boat ramps and construction of new boat ramps and construction of kiosks to provide environmental education to boaters regarding water quality and sustainable practices in coastal areas of Florida.

3.2.6.1.1 Background and Project Description

Escambia County public boat ramps provide local boaters with access to public waterways. This infrastructure provides some of the access for a number of water-dependent recreational activities including fishing, SCUBA diving, water-skiing, and simply cruising local waterways under power or sail. This project would entail repairing an existing boat ramp in Pensacola Bay (Navy Point Park Public Boat Ramp N30-22.8'/W087-16.9') and construction of a new boat ramp facility in Pensacola Bay (Mahogany Mill Public Boat Ramp N30-23.9'/W087-14.9') (Figure 16). The project also includes repairing and modifying an existing boat ramp in Perdido Bay (Galvez Landing Public Boat Ramp N30-18.8'/W087-26.5') and construction of a new boat ramp facility in Perdido Bay (Perdido Public Boat Ramp N30-31.4'/W087-26.7') (Figure 17). Finally, visitor information kiosks would be installed to provide environmental education to boaters regarding water quality and sustainable practices for utilization of marine/estuarine/coastal resources in Florida. The need for enhancements and new ramps at these locations was determined by Escambia County's Marine Advisory Council and was approved by the Board of County Commissioners.

Figure 16. Mahogany Mill public boat ramp design.

The estimated cost for this project is approximately $5,067,255. This cost does not include matching funds provided by local government.

3.2.6.1.2 Selection Criteria

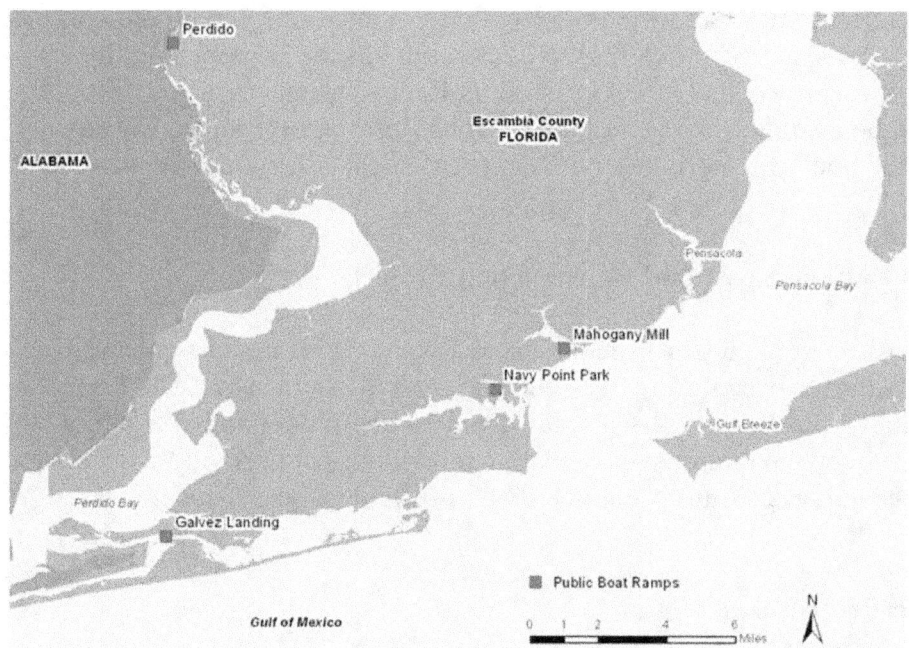

The Florida Boat Ramp Enhancement and Construction Project is intended to improve the quantity and quality of recreational boating in Florida's Pensacola and Perdido Bay systems. Specifically, enhancing public boat ramps would provide local boaters with access to public waterways and water recreational activities (including fishing, diving, water-skiing, SCUBA diving, and cruising).

Figure 17. Florida Boat Ramp Enhancement and Construction Project locations, Escambia County, FL.

This project meets the evaluation criteria for the Framework Agreement and OPA. The project would address the reduced quality and quantity of recreational activities (e.g., boating and fishing) that resulted from natural resource injuries caused by the Spill (*See* CFR § 990.54(a)(2) and Sections 6a-6c of the Early Restoration Framework Agreement) using established techniques. State and local government agencies have successfully completed similar recreational boating projects. Cost estimates are based on similar past projects. As a result, the project is considered feasible, cost effective, and consistent with long-term restoration needs (*See* CFR § 990.54(a)(1),(3),(4) and Sections 6d-6e of the Early Restoration Framework Agreement).

Boat ramp enhancement in Escambia County was suggested as a restoration measure during NOAA's public scoping meetings for the *Deepwater Horizon* PEIS in Florida, submitted as a restoration project on the NOAA website (http://www.gulfspillrestoration.noaa.gov) and submitted to the State of Florida. In addition to meeting the evaluation criteria for the Framework Agreement and OPA, the boat ramp enhancement project meets Florida's criteria that early restoration projects occur in the 8-county panhandle area that deployed boom and was impacted by the Spill, and the project can be completed within 18 to 24 months. Visit the State of Florida's website (http://www.dep.state.fl.us/deepwaterhorizon/projects.htm) to see the 152 panhandle projects (List 2) currently being considered for Early Restoration funding.

3.2.6.1.3   Performance Criteria, Monitoring and Maintenance

Maintenance of boat ramps involves keeping the area clean of debris, emptying trash, repair of onsite facilities, and similar tasks. The first fifteen years of Operation and Maintenance costs would be provided by BP and are included in the total cost of the project, after which maintenance would be completed by Escambia County.

3.2.6.1.4   Offsets

For the purposes of negotiations of Offsets with BP in accordance with the Framework Agreement, the Trustees used monetized estimates of project benefits to estimate Offsets for the Florida Boat Ramp Enhancement and Construction Project, resulting in a monetary value expressed in present value year 2011 dollars. The Trustees considered a number of factors in estimating present value year 2011 dollars, including, but not limited to, initial annual value based on the economic model described in the *Florida Boating Access Facility Inventory and Economic Study* (Florida Fish and Wildlife Conservation Commission, 2009), estimated changes in value over time and expected partial funding from other sources. Total estimated Offsets for the Florida Boat Ramp Enhancement and Construction Project is $10,153,642. These Offsets are reasonable for this resource and this project.

3.2.6.2 Florida (Pensacola Beach) Dune Restoration

Primary vegetated dune habitat located in the Pensacola Beach area of Escambia County and other parts of Florida was injured by exposure to *Deepwater Horizon* oil and/or the extensive use of all-terrain vehicles, heavy equipment and personnel on beaches during response activities undertaken to prevent, minimize and/or remediate oiling. This habitat is located along seaward, frontal dunes, and characterized by a mixture of open sandy areas, grasses and forbs. The vegetative community is typically dominated by plants such as sea oats, panic grass, beach morning-glory, and seashore elder. The Florida (Pensacola Beach) Dune Restoration Project would help restore primary vegetated dune habitat lost due to Spill-related activities.

3.2.6.2.1 Background and Project Description

The goal of this project is to provide early restoration for some of the natural resources that have been injured as a result of the Spill. The project would help restore an area of the beach where oiling and the extensive use of all-terrain vehicles and heavy equipment has inhibited plant growth and prevented the natural seaward expansion of the dunes since June 2010. The primary dunes are the first natural line of defense for coastal Florida to prevent the loss of wildlife habitat and private property due to hurricanes, sea level rise, oil spills, and other threats.

Pensacola Beach is located toward the western end of Santa Rosa Island in Escambia County, Florida. The western boundary of Pensacola Beach lies approximately 7.5 miles east of Pensacola Pass. From that point of origin the project would extend approximately 4.2 miles to the east. This beach segment has been engineered and augmented through two prior nourishment projects.

Approximately 394,240 native plants would be planted approximately 40 feet seaward of the existing primary dunes within designated project areas (Figure 18). Proportions of plants would include approximately 70% sea oats grasses, 20% panic and smooth cord grasses, and 10%

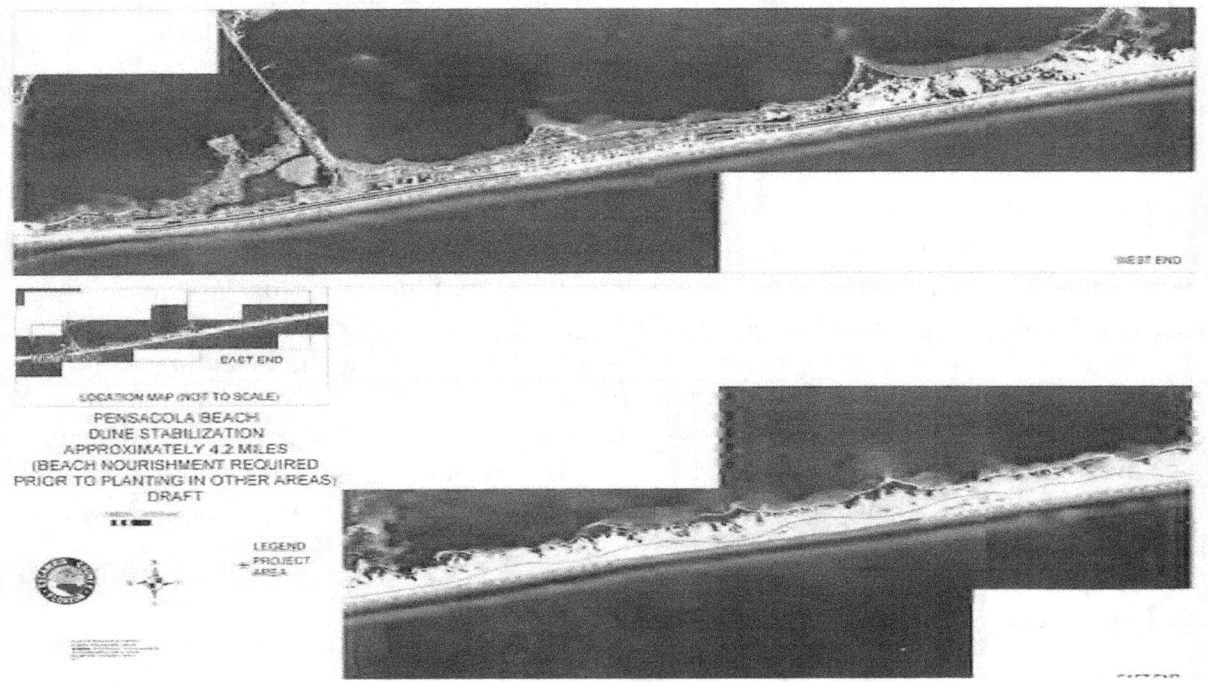

Figure 18. Florida Dune Restoration Project planting areas.

ground cover plants (sea purslane, beach elder, white morning glories and railroad vine) to maximize sand stabilization and limit wind erosion. All plants would be grown from seeds or cuttings from the Alabama or North Florida coast to ensure appropriate genetic stocks are used in the project. Plants would be installed at 18-inch centers and 6 inches deep to ensure that sufficient moisture is available to roots, and properly covered with sand to stabilize and protect the plants. Planting would be targeted for the March – August time frame. Slow release fertilizer would be added during plant installation and plants would be periodically watered, as needed, to facilitate establishment.

The estimated cost for this project is approximately $644,487.

3.2.6.2.2   Selection Criteria

The Florida (Pensacola Beach) Dune Restoration Project would directly restore primary vegetated dune habitat in Florida injured by the Spill through active replacement of plants and dunes. As with the Alabama Dune Restoration Cooperative Project, the project would help prevent erosion by restoring a "living shoreline": a coastline protected by plants and associated dunes rather than hard structures. These natural resources provide habitat to wildlife and increase the storm protection to both habitat and human resources.

This project meets the evaluation criteria for the Framework Agreement and OPA. The project would restore the equivalent of natural resources (vegetated dune habitat) injured by the Spill (See CFR § 990.54(a)(2) and Sections 6a-6c of the Early Restoration Framework Agreement) using established techniques. Trustees and their partners have successfully completed similar dune habitat restoration projects along the northern Gulf coast using these same protocols for decades. Cost estimates are based on similar past projects. As a result, the project is considered feasible, cost effective, and consistent with long-term restoration needs (See CFR § 990.54(a)(1),(3),(4) and Sections 6d-6e of the Early Restoration Framework Agreement).

Dune restoration in Escambia County was suggested as a restoration measure during NOAA's public scoping meetings for the Deepwater Horizon PEIS in Florida, submitted as a restoration project on the NOAA website (http://www.gulfspillrestoration.noaa.gov) and submitted to the State of Florida. In addition to meeting the evaluation criteria for the Framework Agreement and OPA, the Florida dune restoration project meets Florida's criteria that early restoration projects occur in the 8-county panhandle area that deployed boom and was impacted by the Spill and the project can be completed within 18 to 24 months. Visit the State of Florida's website (http://www.dep.state.fl.us/deepwaterhorizon/projects.htm) to see the 152 panhandle projects (List 2) currently being considered for Early Restoration funding.

3.2.6.2.3   Performance Criteria, Monitoring and Maintenance

Large storm events, severe drought and ongoing oil spill cleanup activities could potentially negatively affect the success of plantings in dune habitat restoration. The State or County would monitor plant installations to evaluate project success and recommend maintenance activities for 3-5 years from initial project implementation. County officials would visit project locations on a weekly basis to document survivorship of installed dune plants. Plants that do not survive within the first 90 days after planting would be replaced.

3.2.6.2.4 Offsets

For the purposes of negotiations of Offsets with BP in accordance with the Framework Agreement, the Trustees used Habitat Equivalency Analysis to estimate Offsets provided by the Florida (Pensacola Beach) Dune Restoration Project. Offsets reflect units of discounted service acre years (DSAYs) of primary dune habitat, and would be applied against primary dune habitat along the Florida coast injured by the Spill as determined by the Trustees' total assessment of injury. In estimating DSAYs, the Trustees considered a number of factors, including, but not limited to, benefits of revegetating primary dune habitat, the time period that it would take for revegetated habitat to provide different levels of ecological benefits, estimated project life span and potential impact of hurricanes and drought. Total estimated Offsets for the Florida (Pensacola Beach) Dune Restoration Project is 105 DSAYs. These Offsets are reasonable for this resource and this project.

CHAPTER 4   ENVIRONMENTAL CONSEQUENCES

The Trustees selected the eight Phase I early restoration projects described in Chapter 3 of this ERP/EA. These projects address an array of natural resources and their services injured by the Spill. Each project is located in one of four states, Alabama, Florida, Louisiana, and Mississippi, and in one case includes DOI-managed land. Specifically, the projects are: Marsh Island (Portersville Bay) Marsh Creation; Alabama Dune Restoration Cooperative Project (partially including DOI land); Florida Boat Ramp Enhancement and Construction Project; Florida (Pensacola Beach) Dune Restoration; Mississippi Oyster Cultch Restoration; Mississippi Artificial Reef Habitat; Lake Hermitage Marsh Creation – NRDA Early Restoration Project; and Louisiana Oyster Cultch Project.

Under the National Environmental Policy Act (NEPA), 42 U.S.C. §§ 4321 *et seq.*, federal agencies must consider and disclose the environmental impacts of major federal actions, such as undertakings on federal lands, issuing permits, or providing funding. Federal agencies may categorically exclude certain actions from further NEPA analysis because they characteristically do not have a significant effect on the human environment, individually or cumulatively. An EA is prepared for actions that do not qualify for a categorical exclusion, and is a concise public document that provides information to determine if an action involves significant environmental impacts. If an environmental assessment does not lead to a FONSI and instead identifies a potential for significant environmental impacts, then the agency must prepare an EIS.

The Trustees combined these eight projects into one early restoration plan under OPA rather than preparing eight separate plans. However, for the purposes of NEPA, the Trustees analyzed each project separately. Pursuant to NEPA, this Chapter 4 of the ERP/EA sets forth the purpose and need for each action and analyzes the direct, indirect, and cumulative impacts of the construction and operation (when applicable) of each project and a no action alternative. These analyses resulted in categorical exclusions for two projects and FONSIs for five projects (Appendix E). Further information on the construction and design of the Marsh Island (Portersville Bay) Marsh Creation project will be developed as part of the NEPA approval process.

The eight projects are analyzed in separate NEPA analyses because they each have independent utility. NEPA requires actions that are connected or dependent on other actions to be analyzed together in one NEPA analysis. Actions are considered connected if:

- They automatically trigger other actions which may require an EIS(s).
- They cannot or will not proceed unless other actions are taken previously or simultaneously.
- They are interdependent parts of a larger action and depend on the larger action for their justification.

Proposed projects do not fit the description of connected actions in 40 C.F.R. § 1508.25 requiring analysis through the same environmental analysis. First, to the best of the Trustees' knowledge, none of these projects would automatically trigger other actions which may require an EIS(s). Second, each of these proposed projects[22] represents a whole project and their performance does not depend on the previous or simultaneous performance of any other action. In fact, several of the projects were conceived prior to the Spill and have permits and/or NEPA documentation under separate and unrelated initiatives, but lacked funding for planning and/or implementation. Third, the proposed projects are not an interdependent part of a larger action. Each of these projects are justified and would be undertaken regardless of whether the other proposed projects would be undertaken, and regardless of whether any additional future restoration is undertaken. The Trustees developed, evaluated, and negotiated with BP each of the projects independent from the others. While the Trustees intend to complete one billion dollars in early restoration projects under the Framework Agreement, additional restoration projects are subject to future negotiations. Therefore, each project, including their direct, indirect, and cumulative impacts, has been analyzed separately under NEPA.

Each section in this chapter is organized by project and analyzes the following:
- Aesthetics and Visual Resources
- Air Quality
- Biological Resources
- Cultural Resources
- Geology, Soils and Sediments
- Land Use
- Noise
- Socioeconomics and Environmental Justice
- Public Access/Recreation
- Utilities and Public Services
- Water Resources
- Cumulative Impacts

4.1   Lake Hermitage Marsh Creation – NRDA Early Restoration Project

This proposed NRDA early restoration project involves an additional increment of approximately 100 acres of marsh creation into an existing project known as the "Lake Hermitage Marsh Creation – NRDA Early Restoration Project" that has been developed and is being funded through the Coastal Wetlands Planning, Protection and Restoration Act (CWPPRA) program.

This project is a marsh creation project in an area that was historically marsh but is currently primarily open water located within the Barataria Hydrologic Basin in Plaquemines Parish, Louisiana. For more information, please refer to Section 3.2.2.1.

---

[22] The ERP documents the Trustees' selection of projects. Projects in Chapter 4 are referred to as "proposed projects" because the Trustees analyzed these projects under NEPA prior to the selection of projects under OPA and the Framework Agreement.

NEPA Compliance

The "Final Environmental Assessment, Lake Hermitage Marsh Creation, BA-42" and Finding of No Significant Impact were prepared by the U.S. Fish and Wildlife Service and finalized in November 2011 (Appendix E).

4.2   Louisiana Oyster Cultch Project

Purpose and Need

Louisiana's oyster resources were exposed to oil, dispersant, as well as response activities undertaken to prevent, minimize, or remediate oiling from the Spill. Under OPA, the Trustees act on behalf of the public to restore, rehabilitate, replace, or acquire the equivalent of natural resources injured and associated service losses as a result of the Spill. Under the Framework Agreement, the Trustees have the opportunity to negotiate with BP to fund early restoration projects. The purpose of a Louisiana oyster restoration project implemented under OPA and the Framework Agreement is to begin to restore, rehabilitate, replace or acquire the equivalent of Louisiana's oyster resources injured by the Spill.

General Project Information

Louisiana's oyster resources are among the largest and most valuable in the United States. Habitat exists for oysters throughout many of Louisiana's coastal areas (LDWF, 2010). Throughout coastal Louisiana, the Louisiana Department of Wildlife and Fisheries (LDWF) manages approximately 1.7 million acres of public oyster bottoms, including an estimated area of 38,000 acres of known subtidal reefs (LDWF, 2010). The proposed locations for the Louisiana Oyster Cultch Project include locations within: Chandeleur Sound (cultch locations: Lake Fortuna, S. Black Bay, Drum Bay, 3-Mile Bay), Barataria Bay (cultch location: Hackberry Bay), and Timbalier/Terrebonne Bay (cultch location: Sister Lake) (see Figure 8 in Section 3.2.2.2).

Chandeleur Sound and Breton Sound form part of the Lake Pontchartrain Basin. Together, they comprise more than 500,000 acres. Chandeleur Sound was historically separated from the open waters of the Gulf of Mexico by the Chandeleur Islands and their shallow seagrass beds (Moretzsohn et al., 2011). Average water depths in the Sound are approximately 9 feet; average salinity is 27 ppt. The nearshore areas are comprised of a complex array of bayous, canals, channels, and small embayments (Moretzsohn et al., 2011). The Sound is home to a number of commercially important species, including red drum, spotted seatrout and brown and white shrimp, as well as Federally Endangered species such as the Gulf sturgeon and Kemp's Ridley Sea Turtle (USGS, 2002).

Barataria Bay is located between Bayou Lafourche to the west and the Mississippi River delta to the east; its surface area is estimated at over 400,000 acres (Moretzsohn et al., 2011). Barataria Bay is separated from the open waters of the Gulf of Mexico by a series of barrier islands. Average water depths in the Bay are approximately 6.5 feet; average salinity is 13 ppt (Moretzsohn et al., 2011). Barataria Bay has been designated an estuary of national significance by the EPA National Estuary Program (Moretzsohn et al., 2011). The area includes fresh, intermediate, brackish and saline marshes (CWPPRA, 2011). These habitats provide nursery and breeding grounds for migratory birds and a number of recreationally and commercially important species, including finfish, shellfish, songbirds, ducks and geese (Moretzsohn et al., 2011).

Timbalier/Terrebonne Bay is located between the Atchafalaya River and Bayou Lafourche just west of the Mississippi River Delta. The Timbalier-Terrebonne Bay system includes a complex array of small embayments, bayous, marshlands and islands; it has been designated an estuary of national significance by the EPA National Estuary Program. Average water depths in the Sound are approximately 6.5 feet; average salinity is 18 ppt. The area is home to over 730 bird species, finfish, shellfish, reptiles, amphibians, and mammals (Moretzsohn et al., 2011).

Louisiana is a national leader in the production of commercial oysters, accounting for more than half of oysters landed among the Gulf of Mexico states. The dockside value of landed oysters was over $50 million in 2009 (LDWF, 2010). Public oyster resources are considered the "backbone" of the Louisiana oyster resource – contributing directly to oyster landings and providing a source of seed oysters for transplanting to private leases (LDWF, 2010).

This NRDA early restoration project is comprised of two components. The first component involves the placement of oyster cultch material onto approximately 850 acres of public oyster seed grounds in coastal Louisiana; the second component involves construction of an oyster hatchery facility that would serve to improve existing oyster hatchery operations and provide a supplemental source of oyster larvae and oyster seed. The oyster cultch placement project would place oyster cultch material such as clean oyster shell or other clean hard substrate (i.e., limestone, crushed concrete) onto existing public oyster seed grounds. The hatchery project involves construction of a building that would house aquaculture tanks for oyster broodstock and larvae, and tanks of algae for supplying food for the oyster broodstock and larvae. The new, two-story facility would be approximately 100 ft. X 84 ft. in size, and would be located next to the LDWF Research Laboratory on Grand Isle at a site leased by Louisiana State University. For project details, please refer to Section 3.2.2.2.

4.2.1 Aesthetics and Visual Resources

Affected Environment

Oyster Cultch Placement
The environment to be affected by the proposed project consists of the open water viewshed visible within coastal Louisiana waterbodies.

Oyster Hatchery
The proposed oyster hatchery facility would be located at 133 Port Drive in Grand Isle, Louisiana at a site that is currently leased by Louisiana State University from the Grand Isle Port Commission. The site is currently undeveloped, but is graded and mowed.

Environmental Consequences

Oyster Cultch Placement
The use of barges and large equipment could have a temporary visual impact during the time of project implementation. However, the time needed for the cultch deployment is short and

therefore visual and aesthetic impacts will be short. The vertical profile to be constructed by cultch placement is designed to be below the water surface, comprising less than 10% of the water column depth, and should not be visible from above the water.

Oyster Hatchery

The hatchery facility would be located next to a similar laboratory facility and would not alter the aesthetic and visual character of the area.

No Action

If no activities were to take place, aesthetics and visual resources would not be impacted for either affected area.

4.2.2   Air Quality

Affected Environment

Oyster Cultch and Oyster Hatchery

In November, 2011 air quality within coastal Louisiana was designated by the U.S. EPA as being in attainment with the National Ambient Air Quality Standards, with the exception of the 2008 lead National Ambient Air Quality Standard because three complete years of monitoring data are not yet available to make a final lead attainment designation (see http://www.deq.state.la.us/portal/tabid/112/Default.aspx).

Environmental Consequences

Oyster Cultch Placement

Project implementation will require the use of heavy equipment which could temporarily lead to air pollution due to equipment exhaust. Fine particulate matter associated with the oyster cultch may become airborne during the deployment process. Available best management practices would be employed to prevent, mitigate, and control potential air pollutants during project implementation. Any minor pollution that does occur would be localized and short in duration.

Oyster Hatchery

Air quality may temporarily be impacted during the construction process, due to machinery, equipment, and dust. Available best management practices would be employed to prevent, mitigate, and control potential air pollutants during project construction. After construction, no adverse effects to air quality are anticipated.

No Action

If no action were taken, there would be no impact on air quality.

4.2.3   Biological Resources

Affected Environment

Oyster Cultch Placement

The coastal and nearshore biological resources of Louisiana consist of a diverse group of marine and benthic species and ecologically valuable habitats, including, but not limited to, oyster reefs.

LDWF monitors the size and health of oysters on nearly 1.7 million acres of public grounds. Known subtidal reefs on public oyster grounds are estimated at 38,000 acres, but it is likely that additional acres of reef exist (LDWF, 2010).

Table 3 lists the nine federally listed threatened and endangered species that potentially could occur or pass through the project area.

Table 3: Federally listed threatened and endangered species that could occur in the Project Area.

Common Name	Scientific Name	Status	Parish	Habitat
Piping plover	*Charadrius melodus*	LT	Jefferson, Plaquemines, St. Bernard, Terrebonne	Beaches and mudflats in southeastern coastal areas
Pallid sturgeon	*Scaphirhynchus albus*	LE	Jefferson, St. Bernard	Large rivers of southeastern US; turbid rivers with sandy bottom; in Louisiana, may be seen in Mississippi, Atchafalaya, and Pontchartrain basins
Gulf sturgeon	*Acipenser oxyrhynchus desotoi*	LT	St. Bernard	Estuaries and coastal shelf; spawns in major rivers that empty into the Gulf of Mexico; may be found in large rivers in Pontchartrain basin and adjacent areas
Loggerhead sea turtle	*Caretta caretta*	LT	St. Bernard	Marine deep and shallow water; also inshore areas, bays, salt marshes, ship channels, and mouths of large rivers; in Louisiana, found in Mississippi, Pontchartrain and Barataria Basins
West Indian Manatee	*Trichechus manatus*	LE	Plaquemines, St. Bernard, Terrebonne	Fresh and salt water in large coastal rivers, bays and estuaries.
Kemp's Ridley sea turtle	*Lepidochelyskempii*	LE	All coastal areas	Nearshore waters, estuaries, salt marshes, sandy beaches
Leatherback sea turtle	*Dermochelys coriacea*	LE	All coastal areas	Open ocean and deeper waters of the Gulf and coastal bays; coastal beaches and barrier islands (nesting)
Hawksbill sea turtle	*Eretmochelys imbricata*	LE	All coastal areas	Warm bays and shallow portions of oceans; seagrass beds; estuaries; mainland beaches and islands (nesting)

| Green sea turtle | *Chelonia mydas* | LT | All coastal areas | Sightings in Louisiana are rare; no known nesting sites |

LT = listed threatened, LE = listed endangered, Source: LA Natural Heritage Program Endangered Species by Parish List (http://www.wlf.louisiana.gov/wildlife/species-parish-list?tid=All&type_1=All)

The project area also includes Essential Fish Habitat (EFH) as defined by the Magnuson-Stevens Fishery Conservation and Management Act. EFH encompasses waterbodies, habitats, and substrates necessary for federally and regional fishery management council managed fish to complete various life history stages such as breeding, spawning, feeding or growth and survival to maturity. Table 4 lists the different types of EFH that are associated with the vicinities of the proposed cultch placement locations.

Table 4. Different types of EFH found in the vicinity of proposed cultch placement locations. (The proposed cultch placement locations will not adversely affect EFH).

Proposed Location of Cultch Placement	Essential Fish Habitat Categories in the Vicinity of Proposed Cultch Locations
Sister Lake	Coastal Migratory Pelagics, Red Drum, Reef Fish, Shrimp
Hackberry Bay	Coastal Migratory Pelagics, Red Drum, Reef Fish, Shrimp, Atlantic Sharpnose Shark, Bull Shark, Finetooth Shark, Scalloped Hammerhead Shark, Spinner Shark
S. Black Bay	Coastal Migratory Pelagics, Red Drum, Reef Fish, Shrimp, Atlantic Sharpnose Shark, Blacktip Shark, Bull Shark, Finetooth Shark, Scalloped Hammerhead Shark
Lake Fortuna	Coastal Migratory Pelagics, Red Drum, Reef Fish, Shrimp, Atlantic Sharpnose Shark, Blacktip Shark, Scalloped Hammerhead Shark
Drum Bay	Coastal Migratory Pelagics, Red Drum, Reef Fish, Shrimp, Atlantic Sharpnose Shark, Blacktip Shark, Bull Shark, Finetooth Shark, Scalloped Hammerhead Shark
3-Mile Bay	Coastal Migratory Pelagics, Red Drum, Reef Fish, Shrimp, Atlantic Sharpnose Shark, Blacktip Shark, Bull Shark, Finetooth Shark, Scalloped Hammerhead Shark

Oyster Hatchery

The proposed site is located in an area with existing similar facilities. Wildlife adapted to human presence (e.g., raccoons, birds, etc.) may be found in the area. Vegetation is either landscaped, or weedy. No noxious weeds or invasive species are known to occur in the proposed project area. Piping plover is the only federally listed threatened or endangered species found in terrestrial habitats in Jefferson Parish, where Grand Isle is located. However, the FWS has evaluated whether this project would affect the piping plover under the Endangered Species Act and has concluded that this species is not found in the proposed project area, and therefore will not be affected by the project.

Environmental Consequences

Oyster Cultch Placement

Short-term disturbances to water column and benthic organisms may occur when the project is implemented. As cultch material is deployed, any planktonic organisms could be displaced due to the falling material. As the material settles to the seafloor, there would be displacement of and loss of infauna and some epifauna within the area of deployment. Turbidity levels may be locally increased in the area where shell cultch is deployed but would be of short (hours) duration. Some epifaunal organisms are mobile enough to move away from the affected area before the material settles. Although there may be temporary impacts to the existing benthic community as a result of project implementation, the completed project would result in improved oyster secondary production. Recent oyster cultch placement projects in Louisiana have been permitted under the New Orleans District Corps of Engineers Programmatic General Permit (PGP) for the Louisiana Coastal Zone. Louisiana intends to apply for authorization for the proposed cultch placement project under the PGP. Additionally, the Louisiana Department of Natural Resources (LDNR) evaluated the proposed project and determined the project to be broadly consistent with the Louisiana Coastal Resource Program (LCRP). LDNR will provide a final determination upon receipt of the final consistency determination or Coastal Use Permit application for the project.[23] The Trustees would follow best management practices to avoid affecting existing environmentally sensitive areas for cultch placement. Examples of sensitive areas include viable productive oyster reefs, emergent and submerged aquatic vegetation, and other live bottom communities.

The FWS evaluated whether this project would affect piping plover and pallid sturgeon under Section 7 of the Endangered Species Act, and has concluded that these species are not found in the proposed project area, and therefore will not be affected by the project. The FWS has also evaluated whether this project would affect the West Indian manatee and has concluded that the project may affect, but is not likely to adversely affect the West Indian manatee. The National Marine Fisheries Service has evaluated whether the project may affect the Gulf Sturgeon, the Leatherback sea turtle, the Hawksbill sea turtle, the Loggerhead sea turtle, the Kemp's Ridley sea turtle, and the Green sea turtle, and has concluded that the project is not likely to adversely affect these species. ESA consultation would also be completed as necessary during permitting processes. The National Marine Fisheries Service has evaluated the proposed cultch placement and oyster hatchery project and concluded that the project would not adversely affect EFH, and overall, would likely benefit federally managed fishery species in areas where proposed project locations may affect EFH areas (see Table 4).

The threatened Gulf sturgeon (*Acipenser oxyrinchus desotoi*) is an anadromous fish that overwinters in the Gulf of Mexico and adjacent estuaries and bays. This species utilizes soft sedimentary substrate habitats (sand, silt, clay) for foraging. Populations of Gulf sturgeon are found in the Pearl River system (including the Pearl and Bogue Chitto Rivers) in Louisiana (Kirk, 2008). The Pearl River system and coastal waters extending from the outflow of the Pearl River toward Mississippi are included within the designated critical habitat areas for the Gulf sturgeon (68 FR 13370; see: http://www.nmfs.noaa.gov/pr/pdfs/criticalhabitat/gulfsturgeon.pdf).

---

[23] Coastal Zone Management Act compliance documentation for projects proposed in the ERP/EA will be available in the Administrative Record.

The closest proposed cultch placement location to these areas is 3-Mile Bay, and the National Marine Fisheries Service has concluded under section 7 of the Endangered Species Act that the project is not located in Gulf sturgeon critical habitat. The National Marine Fisheries Service also determined that this project is not likely to adversely affect the Gulf Sturgeon.

Ross et al. (2008) performed telemetry studies which indicated that Gulf sturgeon were present in Mississippi Sound habitats from October through March. In addition, these telemetry studies showed that once Gulf sturgeon leave the freshwater riverine spawning habitats they typically are found in the shallow water habitats of the barrier island passes with no occurrences in the nearshore habitats of the proposed project. This suggests that sturgeon presence in the project area may only occur during seasonal migrations to barrier island shallow waters. The foraging habitat of sturgeon is mainly soft, sandy substrate not the hard substrate of existing oyster reef. A limited amount of soft substrate, and sturgeon foraging habitat, could potentially be lost during and following deployment. Based on currently available information regarding the life cycle of the Gulf sturgeon and the location and timing of cultch deployment, it is unlikely that the Gulf sturgeon would be adversely impacted by the proposed project.

Oyster Hatchery
The FWS evaluated whether this project would affect piping plover and pallid sturgeon and has concluded that these species are not found in the proposed project area, and therefore will not be affected by the project. The FWS also evaluated whether this project would affect the West Indian manatee has concluded that the project may affect, but is not likely to adversely affect the West Indian manatee. The National Marine Fisheries Service has evaluated whether the project may affect the Gulf Sturgeon, the Leatherback sea turtle, the Hawksbill sea turtle, the Loggerhead sea turtle, the Kemp's Ridley sea turtle, and the Green sea turtle, and has concluded that the project is not likely to adversely affect these species. ESA consultation would also be completed as necessary during permitting processes. The National Marine Fisheries Service has evaluated the proposed cultch placement and oyster hatchery project and concluded that the project would not adversely affect EFH, and overall, would likely benefit federally managed fishery species in areas where proposed project locations may affect EFH areas (see Table 4).

Construction
Construction of a facility at this location would likely not impact any other threatened and endangered species or wildlife populations in general. The hatchery site is currently undeveloped, but is graded and mowed. Urban wildlife would adapt to the additional disturbances created by construction and operational activities.

Operation
The provision of oyster larvae and oyster seed are not expected to have any adverse impact on biological resources because the oyster hatchery uses native broodstock that would not affect the genetic characteristics of the oyster population. The project would result in benefits by improving the success rate of the oyster cultch placement component and increasing oyster production. If hatchery activities were not undertaken to supplement cultch placement, oyster production achieved under the oyster cultch placement component would likely be reduced.

No Action

Currently degraded habitat conditions and reduced oyster productivity would remain at the cultch placement sites. No impacts to currently existing biological resources at the hatchery site would occur.

4.2.4 Cultural Resources

Affected Environment

Oyster Cultch Placement

The area of potential effect (APE) for reviews under Section 106 of the National Historic Preservation Act includes the areas of direct and indirect impact. For this component of the proposed project, it consists of the footprint of the oyster cultch placement. Cultural resources could potentially be affected in the project area; however, no known cultural resources, including shipwrecks, are located in the project area as evidenced from recent side-scan sonar surveys of the water bottoms.

Oyster Hatchery

No known cultural resources are located within the project area. The soil at the construction site consists of dredge spoils.

Environmental Consequences

Oyster Cultch Placement

Louisiana intends to seek authorization for the proposed oyster cultch placement under the New Orleans District Corps of Engineers PGP for the Louisiana Coastal Zone. The PGP includes an assurance from the New Orleans District Army Corps of Engineers that all projects eligible for the PGP would be screened for impacts to historic or cultural resources from information on file with the New Orleans District. A complete review of this project under Section 106 of the National Historic Preservation Act will be completed prior to project implementation. Any culturally or historically important resources will be avoided during site selection. This project would be implemented in accordance with all applicable laws and regulations concerning the protection of cultural and historic resources.

Oyster Hatchery

A complete review of this project under Section 106 of the National Historic Preservation Act will be completed prior to project implementation. This project would be implemented in accordance with all applicable laws and regulations concerning the protection of cultural and historic resources.

No Action

Cultural resources would not be impacted if the project were not implemented.

4.2.5 Geology, Soils, and Sediments

Affected Environment

Oyster Cultch Placement

The substrates in coastal Louisiana include soft sediments and hard reef substrates. Locations proposed for oyster cultch placement may include areas where hard reef substrates have existed in the past.

Oyster Hatchery

Soils on Grand Isle are typical of those associated with Holocene coastal marshes. The surface of this area is primarily Mississippi River clay, silt and fine sand, including recent alluvial material and Pleistocene-age marine sediments (Weindorf, 2008). The soil at the site consists of dredge spoils.

Environmental Consequences

Oyster Cultch Placement

There should be minimal adverse impacts to geology, soils, or sediments. This action could potentially replace a limited amount of soft sedimentary substrates with hard substrates. The project would create low profile alterations above the substrate to localized areas of the seafloor. The low profiles of the deployed cultch areas are intentional so as to minimize impacts from currents.

Oyster Hatchery

Except for the direct footprint of the building, the proposed hatchery construction component would not have adverse impacts to soils in the surrounding environment. Geology and sediments would not be impacted.

No Action

There would be no changes to existing geology, soils, and sediment.

4.2.6   Land Use

Affected Environment

Oyster Cultch Placement

The proposed project areas consist of open water within coastal Louisiana, and would not include terrestrial or shoreline areas.

Oyster Hatchery

The hatchery would be built in an area already occupied by marine laboratory research facilities.

Environmental Consequences

Oyster Cultch Placement

Implementation of the project would not disrupt existing land uses, shoreline areas, or wetlands. LDNR has evaluated the proposed project and determined the project to be broadly consistent with the LCRP. LDNR will provide a final determination upon receipt of the final consistency determination or Coastal Use Permit application for the project.

Oyster Hatchery

Construction of an oyster hatchery would have no effect on current land use.

No Action

If no action were taken, there would be no impact to land use.

4.2.7 Noise

Affected Environment

Oyster Cultch Placement

The current noise levels are minimal on the open water of the proposed project areas.

Oyster Hatchery

The current noise levels are typical for developed areas in a town with a small population of approximately 1,500 individuals.

Environmental Consequences

Oyster Cultch Placement

This project requires the use of heavy equipment and barges for implementation which would emit noise. Wildlife and humans in the area could be impacted. Noise levels above the existing background levels will be limited to the short duration of cultch deployment.

Oyster Hatchery

There may be a temporary noise impact during the construction process. After the construction is completed there should be no significant increase in the amount or degree of noise.

No Action

There would be no changes in noise conditions.

4.2.8   Socioeconomics and Environmental Justice

Affected Environment

Oyster Cultch Placement

Louisiana is a national leader in oyster production. The combination of public grounds and private leases produces an annual dockside value in excess of $35 million. Louisiana accounted for an average of 34% of the nation's oyster landings from 1998-2008. Among Gulf of Mexico states, Louisiana consistently ranks #1 in landings, accounting for over 50% of oyster landed. Louisiana was the top producer in 2008 with approximately 12.8 million pounds of oysters. In 2009 the dockside value of oysters was over $50 million, the highest ever (LDWF, 2010). This was a result of 14,870,438 million pounds of meat, the second-highest on record (NOAA Fisheries, 2011b). Nearly 90% of public ground oysters harvested in 2008/2009 were harvested from the Louisiana portion of the Mississippi Sound, Lake Borgne, Chandeleur Sound, and the area south of the Mississippi River Gulf Outlet out to the Breton National Wildlife Refuge (LDWF, 2009). Over 75% of public ground oysters harvested in 2010 came from these same areas (LDWF, 2010).

Oyster Hatchery

According to U.S. Census Bureau statistics, in 2000, Grand Isle had a population of 1,541 and a median household income of $33,548 which was below the national median income. In addition, 39% of families were considered to be below the poverty level.

Environmental Consequences

Oyster Cultch Placement

There should be no adverse social, economic, health, or environmental impacts to local communities due to this project. Development of approximately 850 acres of oyster cultch would enhance existing Louisiana oyster management efforts and result in an increase in harvestable oysters. Furthermore, the project would not have a disproportionate effect on any particular group of people or individuals. In fact, development of additional oyster harvest opportunities would provide greater economic and commercial resources for local citizens and local businesses. The project would not have a disproportionate effect on low income or minority populations.

Oyster Hatchery

The hatchery project would have positive impacts on Louisiana's coastal economy by increasing the success of oyster cultch placement through provision of oyster larvae and seed. In addition, construction of the oyster hatchery building and operation of the oyster hatchery would provide greater economic and commercial resources for local citizens and local businesses due to the jobs and expenditures associated with construction and operations. The project would not have a disproportionate effect on low income or minority populations.

No Action

Socioeconomics and environmental justice would not be impacted if the project were not implemented.

4.2.9   Public Access/Recreation

Affected Environment

Oyster Cultch Placement

Louisiana's oyster resources are managed as a combination of public oyster grounds and private leases. The project area would yield a source of seed oysters that can be transplanted to private leases and also yield a supply of harvestable (sack-sized) oysters that may be harvested by recreational or commercial fishermen. The Louisiana Wildlife and Fisheries Commission determines which areas are open for harvest as well as the season opening and closing dates. Public access to the project areas is available for commercial and recreational use.

Oyster Hatchery

The location where the hatchery would be located is an undeveloped, open lot adjacent to another research facility. The site is leased by Louisiana State University from the Grand Isle Port Commission. The public does not currently have access to this open lot and there is no recreation associated with the location.

Environmental Consequences

Oyster Cultch Placement

During placement of cultch material, public access to and recreation within the deployment area may be restricted or limited at times. After cultch placement, seed-sized oysters may be removed from public seed grounds in as little as 4 months after the process of successful spat set. Oysters require approximately two to three years in Louisiana to develop into harvestable size (sack-sized oysters) that would be available for recreational or commercial harvest. Restoration of approximately 850 acres of oyster cultch areas would result in increased public access to the oyster resource.

Oyster Hatchery

Public access to the oyster hatchery building would be controlled by Louisiana State University and the LDWF. Tours and educational outreach events would be offered to the public on a periodic basis, resulting in additional educational benefits to the community. Increasing the success of oyster cultch placement would result in increased public access to the oyster resource. The oyster hatchery would have no other impacts on public access or recreation.

No Action

There would be no change to public access or recreation.

4.2.10 Utilities and Public Services

Affected Environment

Oyster Cultch Placement

Potentially existing utilities or public services within the underwater area of the project are buried beneath the sediment.

Oyster Hatchery

The newly constructed hatchery facility would include a water intake/outfall and filtration system, utilities, and public services

Environmental Consequences

Oyster Cultch Placement

Deployment of cultch material would not disturb any potentially existing utilities or public services in the proposed area as they are buried into the sediment.

Oyster Hatchery

Construction and operation of the oyster hatchery is not expected to have substantial impacts on utilities and public services, including wastewater treatment, and is similar to what is currently used by the adjacent LDWF Research Lab that houses temporary hatchery operations.

No Action

There would be no changes to utilities or public services.

4.2.11 Water Resources

Affected Environment

Oyster Cultch Placement and Oyster Hatchery

Louisiana's water resources consist of wetlands, shorelines, bays, intertidal and subtidal areas, and open water habitat. The project areas border the Mississippi River Delta and are located within several coastal Louisiana basins (including Atchafalaya, Terrebonne, Barataria, Breton Sound, and Pontchartrain).

Environmental Consequences

Oyster Cultch Placement

Temporary sediment and water quality impacts could occur with project implementation. Deployment of cultch material could cause disturbance to bed sediment that could increase turbidity and suspended sediment concentrations in the water column of the deployment area. However, any potential water quality impacts would be minor and localized, lasting several hours to several days at most. Louisiana intends to seek authorization for the proposed oyster cultch placement under the New Orleans District Corps of Engineers PGP for the Louisiana Coastal Zone. This PGP covers Clean Water Act permitting for oyster cultch placement in Louisiana. For oyster cultch placement, the PGP has blanket Louisiana Department of Environmental Quality Water Quality certification.

Oyster Hatchery

Temporary sediment and water quality impacts could occur with project construction, due to erosion or run-off from the project area. However, any potential water quality impacts would be minor and localized to the period of construction. The hatchery system would use the water from, and replace the water to, the bay immediately adjacent to the hatchery facility. The planned capacity for the water system is approximately 8,000 gallons per day from April-October, the months when the hatchery would operate. The hatchery includes a water filtration system. The only addition to the water in the hatchery system is algae, which is taken up by the oyster larvae and broodstock, resulting in no adverse impacts to water quality. In fact, because oysters are filter feeders, the hatchery would likely improve water as water passes through the system.

No Action

If no action were taken, there would be no impact on water resources.

4.2.12 Cumulative Impacts – Louisiana Oyster Cultch Project

LDWF has placed over 1.5 million cubic yards of cultch material on nearly 30,000 acres of water bottoms within Louisiana's public oyster areas since 1917 (LDWF, 2008). Deployment of oyster cultch materials would occur within designated public oyster areas in Louisiana state waters and would be part of a long series of oyster ground rehabilitation efforts that have been undertaken by LDWF. This project also includes hatchery improvements to help facilitate and expedite success of the cultch placement. The construction of an oyster hatchery facility that would benefit the existing Sea Grant oyster hatchery located at the LDWF facility on Grand Isle, Louisiana, and improve the ability of the hatchery to produce oyster larvae that can be broadcast onto suitable coastal areas (i.e., cultched areas), or encouraged to settle (set) onto small pieces of shell in the hatchery. The project would provide additional oyster production in the areas that receive cultch placements and increase oyster harvesting opportunities for both seed-sized and sack-sized oysters. At the same time, finfishing in the area would not be impeded by this project. The project is consistent with the goals of Louisiana to restore and enhance its oyster grounds in coastal waters. The oyster cultch placements and hatchery activities are both consistent with ongoing activities of Louisiana. Although this project may have potential short-term negative effects, on balance, the Louisiana Oyster Cultch Project has positive effects that are consistent with long-term planning goals, and benefit the Louisiana coastal environment, including positive impacts on Louisiana's coastal economy. Additionally, all effects are relatively local and geographically disparate.

No Action

There would be no cumulative impacts under the No Action alternative.

4.2.13 Summary

Overall, this project would enhance Louisiana's oyster productivity. The beneficial ecological impacts are expected to far outweigh any short-term, adverse impacts from deployment of cultch material and/or the construction of the hatchery facility. The Trustees believe that the proposed project will enhance oyster productivity within coastal Louisiana.

4.3   Mississippi Oyster Cultch Restoration

Purpose and Need

Mississippi's oyster resources were exposed to oil, dispersant, as well as response activities undertaken to prevent, minimize, or remediate oiling from the Spill. Under OPA, the Trustees act on behalf of the public to restore, rehabilitate, replace, or acquire the equivalent of natural resources injured and associated service losses as a result of the Spill. Under the Framework Agreement, the Trustees have the opportunity to negotiate with BP to fund early restoration projects. The purpose of a Mississippi oyster cultch restoration project implemented under OPA and the Framework Agreement is to begin to restore, rehabilitate, replace or acquire the equivalent of Mississippi's oyster resources.

General Project Information

The Mississippi Sound extends along the southern coasts of Mississippi and Alabama. The Sound is separated from the Gulf of Mexico by several narrow barrier islands and sand bars (including Cat Island, Ship Island, Horn Island, and Petit Bois Island) which provide dynamic and diverse habitats especially for over 300 species of migratory or permanent resident bird species (USACE, 2009). Along the Mississippi Sound, there are numerous coastal bays including St. Louis Bay, Biloxi Bay, Pascagoula Bay and Grand Bay. Coastal wetlands within the Sound include swamps, tidal flats, brackish and salt water marshes, and bayous. Expansive marsh systems include the Grand Bay marshes and the Pascagoula River marsh system to the east of the Sound, and the Hancock County marshes in the west. These are rich in wildlife resources and provide nesting grounds and important stopovers for waterfowl and migratory birds, as well as spawning areas and valuable habitats for commercial and recreational fish.

The Mississippi Sound is shallow with water depths generally not exceeding 20 feet. Water is exchanged with the Gulf of Mexico through the openings between the barrier islands. Its partially protected nature and the influx of riverine freshwater create a salinity gradient within the Sound (Priddy et al., 1955). This delicate mix of fresh and salt water provides a suitable habitat for oysters, shrimp, and other fisheries. Christmas and Waller (1973) reported 138 fish species in 98 genera and 52 families taken from areas across Mississippi Sound. Vittor and Associates (1982) identified over 437 taxa of macrofauna from the Sound with densities varying from approximately 1,200 to 38,900 individuals/yard^2. In addition, there is a diverse, but not commercially relevant community of crustaceans in the Sound and adjacent waters.

Oysters grow well in areas with fluctuating salinities within their normal ranges (such as in Mississippi Sound), compared to areas with constant salinity (Pierce and Conover, 1954). Oyster reefs of commercial importance are subtidal and form aggregates that cover thousands of acres of the Mississippi Sound. The State of Mississippi's 17 oyster reefs are managed by the Department of Marine Resources (MDMR). Approximately 97% of the commercially harvested oysters in Mississippi come from reefs in the western part of the Mississippi Sound, primarily from Pass Marianne, Telegraph and Pass Christian reefs.

The highly productive Mississippi Sound including its coastal wetlands (e.g., St. Louis Bay, Biloxi Bay, Pascagoula Bay, and the tidal Pascagoula River) supports the commercial fishing industry in the State of Mississippi. A study by Mississippi State University's Coastal Research and Extension Center reported the total economic impact of the Mississippi seafood industry as $489 million annually, including $256 million in income and about 28,000 man-years of employment (Posadas, 2001).

The project consists of the restoration of approximately 1,430 acres of oyster cultch areas in the marine waters of the State of Mississippi. Oyster cultch material such as clean oyster shell or other clean hard substrate (limestone, crushed concrete) would be placed within the footprint of existing oyster cultch areas. No facilities would be constructed as part of this project. For project details, please refer to Section 3.2.3.1.

4.3.1 Aesthetics and Visual Resources

Affected Environment

The environment to be affected by the proposed project consists of the open water viewshed visible within Mississippi Sound, bays, and tidal waterbodies.

Environmental Consequences

The use of barges and large equipment could have a temporary visual impact during the time of project implementation. However, the time needed for the cultch deployment would be short and therefore visual and aesthetic impacts would be short. The placed cultch material would remain under the water surface at all times.

No Action

Aesthetics and visual resources would not be impacted under the No Action alternative.

4.3.2 Air Quality

Affected Environment

The air quality within coastal Mississippi is in attainment with the National Ambient Air Quality Standards (MDEQ, 2010).

Environmental Consequences

Project implementation would require the use of heavy equipment which could temporarily lead to air pollution from equipment exhaust. Some fine particulate matter (dust) associated with the oyster cultch may become airborne during the deployment process. No air quality permits are required for this type of project and Hancock and Harrison counties anticipate no violations of state air quality standards are expected. Available best management practices would be employed to prevent, mitigate, and control potential air pollutants during project implementation. Any potential minor impacts would be localized and short in duration.

No Action

There would be no change in air quality. Hancock and Harrison Counties currently in attainment for state air quality standards.

4.3.3 Biological Resources

Affected Environment

The coastal and nearshore biological resources of Mississippi consist of a diverse group of marine and benthic species and ecologically valuable habitats, including, but not limited to, oyster reefs. The oyster reefs are subtidal and form aggregates that cover approximately 12,000 acres of the Mississippi Sound.

Although coastal Mississippi harbors a number of federally-listed threatened, endangered, or candidate species not all of these typically occur in the nearshore habitat of the project area. Table 5 lists the federal and state listed threatened and endangered species that potentially could occur in the project area. The listed least tern and piping plover use beach, mudflat, and riverine habitats not the nearshore habitat of the project area. In addition, Table B-1 (Appendix B) lists several whale species that are federally listed as threatened or endangered although these likely do not occur in the project area. The green sea turtle, hawksbill sea turtle, Kemp's ridley sea turtle, leatherback sea turtle, loggerhead sea turtle, and West Indiana manatee, do not have more than a transient occurrence, if any, with the proposed project area.

Table 5. Federal and state listed threatened and endangered species that potentially could occur or pass through the Mississippi Oyster Cultch Restoration project area.

Common Name	Scientific Name	Federal Status	State Status	County	Habitat
Green sea turtle	*Chelonia mydas*	LT	LE	Hancock, Harrison, Jackson	Shallow coastal waters with SAV and algae, nests on open beaches.
Gulf sturgeon	*Acipenser oxyrhynchus desotoi*	LT	LE	Hancock, Harrison, Jackson	Migrates from large coastal rivers to coastal bays and estuaries
Hawksbill sea turtle	*Eretmochelys imbricata*	LE	LE	Hancock, Harrison, Jackson	Coral reefs, open ocean, bays, estuaries
Kemp's ridley sea turtle	*Lepidochelys kempii*	LE	LE	Hancock, Harrison, Jackson	Nearshore and inshore coastal waters, often in salt marshes
Leatherback sea turtle	*Dermochelys coriacea*	LE	LE	Hancock, Harrison, Jackson	Open ocean, coastal waters
Loggerhead sea turtle	*Caretta caretta*	LT	LE	Hancock, Harrison, Jackson	Open ocean; also inshore areas, bays, salt marshes, ship channels, and mouths of large rivers
West Indian Manatee	*Trichechus manatus*	LE	LE	Hancock, Harrison, Jackson	Fresh and salt water in large coastal rivers, bays and estuaries.

LT = listed threatened, LE = listed endangered
Source: Mann, 2000.

Environmental Consequences

Short-term disturbances to water column and benthic organisms may occur when the project is implemented. The turbidity in the water may temporarily (hours) increase during deployment. The deployed material is expected to displace or cover some infauna and epifauna. However, many epifaunal organisms are mobile and would be minimally affected by the settling material. Biological impacts would be temporary. Overall, the completed project would result in an improved benthic and marine ecosystem especially for oysters. All effort would be made for cultch placement to avoid existing environmentally sensitive areas such as viable productive oyster reefs, emergent and submerged aquatic vegetation, and other live bottom communities.

The Magnuson-Stevens Fishery Conservation and Management Act applies to activities in essential fish habitat (EFH). EFH protection is provided for federally and regionally managed fisheries. EFH encompasses waterbodies, habitats, and substrates that are necessary for fish to complete various life history stages such as breeding, spawning, feeding or growth and survival to maturity. Within the proposed project areas, habitat that falls within this designation includes the water column and both hard and soft substrates (silt, clay, sand, rock, and shell). The threatened Gulf sturgeon (*Acipenser oxyrinchus desotoi*) is an anadromous fish that overwinters in the Gulf of Mexico and adjacent estuaries and bays. This species utilizes soft sedimentary substrate habitats (sand, silt, clay) for foraging.

The National Marine Fisheries Service evaluated the proposed cultch placement project and concluded that the project would not adversely affect EFH, and overall, would likely benefit federally managed fishery species.

Deployment of oyster cultch occurs during the spring and fall. Ross et al. (2008) performed telemetry studies which indicated that Gulf sturgeon were present in Mississippi Sound habitats from October through March although primarily November through March. Therefore, sturgeon would not be present in the proposed project area during time of deployment. In addition, the telemetry study showed that once Gulf sturgeon leave the freshwater riverine spawning habitats they typically are found in the shallow water habitats of the barrier island passes (Figure 19) with no occurrences in the nearshore habitats of the proposed project. This suggests that sturgeon presence in the nearshore environments is minimal, sporadic, and only occurs during seasonal migrations to barrier island shallow waters. Lastly, the foraging habitat of sturgeon is mainly soft, sandy substrate not the hard substrate of existing oyster reef. Although the proposed project would only place cultch material on existing reef footprints a limited amount of soft substrate, and sturgeon foraging habitat, could potentially be lost during and following deployment. Therefore, due to the life cycle of the Gulf sturgeon, its preferred foraging habitat, and the location and timing of cultch deployment it is likely that the Gulf sturgeon would not be impacted or would only be minimally impacted by the proposed project.

The National Marine Fisheries Service evaluated whether the project may affect the Gulf Sturgeon and it critical habitat under Section 7 of the Endangered Species Act, and has concluded the project is not likely to adversely affect the Gulf sturgeon or its critical habitat. The National Marine Fisheries Service has evaluated whether the project may affect the Leatherback sea turtle, the Loggerhead sea turtle, the Kemp's Ridley sea turtle, the Green sea turtle, and the Hawksbill sea turtle has concluded that the project is not likely to adversely affect those species. The FWS evaluated whether this project would affect the West Indian manatee under Section 7 of the Endangered Species Act, and has concluded that the project may affect, but is not likely to adversely affect the West Indian manatee.

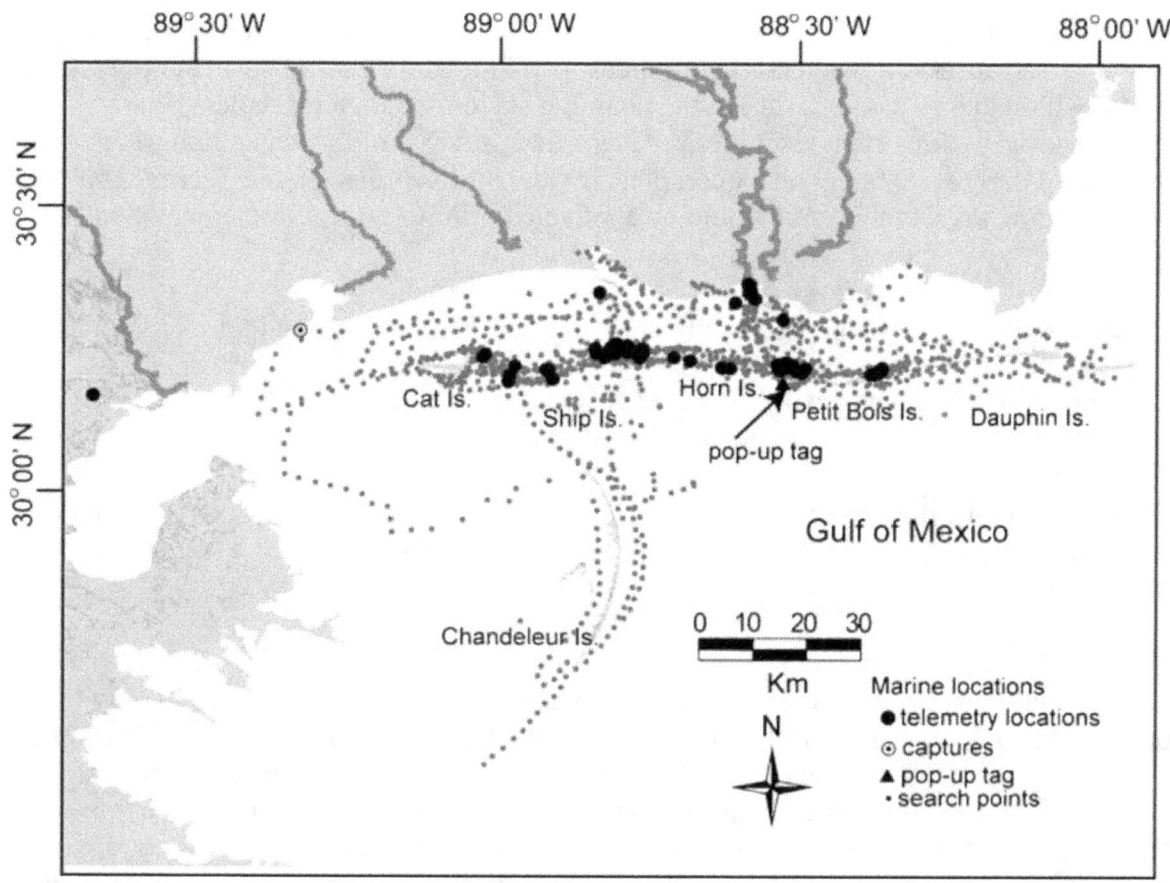

Figure 19. Locations of Gulf sturgeon in the marine environment (large black circles; 1999–2004) and telemetry stations (small gray circles; 2000–2004) by MS personnel. From Ross et al. (2008).

No Action
Currently degraded habitat conditions and reduced oyster productivity would remain.

4.3.4   Cultural Resources

Affected Environment

The area of potential effect (APE) used for reviews under Section 106 of the National Historic Preservation Act includes the areas of direct and indirect impact. For this proposed project it consists of the footprint of the oyster cultch placement. Shipwrecks and their associated artifacts are historical cultural resources that could potentially be affected in the project area. In addition, some locations within Mississippi Sound could contain submerged midden sites (Lewis, 2000).

Environmental Consequences

National Historic Preservation Act Section 106 was considered during the USACE Section 10 permitting process. No shipwrecks or other cultural resources are known to exist in the project area. Consultation with the Mississippi Department of Archives and History (MDAH) was initiated to determine the presence or absence of historic, archaeological, or culturally significant

73

sites (December 1, 2011; MDAH Project Log #12-006-11). MDAH determined that the project is unlikely to adversely affect known historic resources. In addition, a sidescan sonar survey would be completed within the project area during the planning stage for cultch placement. If any culturally or historically important resources are identified during project preparations, such sites would be avoided during site selection. A complete review of this project under Section 106 of the National Historic Preservation Act would be completed prior to project implementation.

No Action

Cultural resources would not be impacted if the project were not implemented.

4.3.5   Geology, Soils, and Sediments

Affected Environment

The proposed project would be implemented within existing oyster reefs which consist primarily of hard reef substrate of shells, limestone, or concrete as well as a very limited amount of soft sediments such as sand, silt, or clay.

Environmental Consequences

There should be minimal impacts to geology, soils, or sediments. Oyster cultch would only be placed on existing oyster reef footprints. This action would mainly cover existing hard substrates although it could potentially replace a limited amount of soft sedimentary substrates with hard substrates. The project would create low profile alterations approximately 1 inch above the substrate to localized areas of the seafloor. The low profiles of the deployed cultch areas are intentional to minimize displacement by currents. In fact, oyster cultch would assist in stabilizing the sea floor during storm events and reduce the mobilization of sediment.

No Action

There would be no changes to existing geology, soils, and sediment.

4.3.6   Land Use

Affected Environment

The proposed project areas consist of open water within Mississippi Sound, and would not include terrestrial or shoreline areas.

Environmental Consequences

Implementation of the project would not disrupt existing land uses, shoreline areas, or wetlands. However, the project would be set up to be consistent with the coast wetlands use designations set forth in the Mississippi Coastal Program and any other applicable local zoning requirements.

No Action

There would be no impact to existing land uses.

4.3.7   Noise

Affected Environment

Noise in the planned deployment areas would be limited to occasional vessel traffic.

Environmental Consequences

This project requires the use of heavy equipment, tug boats, and barges for implementation which would emit noise. Wildlife and humans in the area could be impacted. Noise above the existing background levels would be limited to the short duration of cultch deployment.

No Action

There would be no changes in noise conditions.

4.3.8   Socioeconomics and Environmental Justice

Affected Environment

In 2009, the commercial fishing industry provided approximately 1,200 jobs and generated $61 million in sales and $19 million in personal income (NOAA Fisheries, 2011a). The shellfish fishing sector provided approximately 1,100 jobs (Posadas and Posadas, 2011). From 2007 to 2009, approximately 5.1 million pounds of oysters were commercially landed, generating $13.8 million in income (NOAA Fisheries, 2011b). Approximately 97% of the commercially harvested oysters in Mississippi come from the reefs in western Mississippi Sound, primarily from Pass Marianne, Telegraph and Pass Christian reefs.

Environmental Consequences

There are no anticipated adverse social, economic, health, or environmental impacts to local communities due to this project. Development of 1,430 acres of oyster cultch would enhance existing Mississippi oyster management efforts and result in an increase in harvestable oyster areas. In fact, the project development of oyster harvest opportunities would provide greater economic and commercial resources for local citizens and local businesses due to the enhanced harvesting opportunities.

The project would not have a disproportionate effect on any particular group of people or individuals, including low income or minority populations.

No Action

Socioeconomic conditions and environmental justice would not be impacted if the project were not implemented.

4.3.9   Public Access/Recreation

Affected Environment

Oyster harvest is open to commercial fishing in Mississippi. In addition, Mississippi has a very limited private lease program. For oyster harvesting, MDMR determines harvest area openings and closings and the length of the harvest season.

Environmental Consequences

During placement of oyster cultch, public access to the deployment area would be temporarily restricted. However, the deployment time in any given area is very short and therefore impacts to public access areas in the area are considered minimal.

No Action

There would be no change to public access or recreation.

4.3.10 Utilities and Public Services

Affected Environment

Potentially existing utilities or public services within the underwater area of the project are expected to be buried beneath the sediment.

Environmental Consequences

Deployment of cultch material would not disturb any potentially existing utilities or public services in the proposed area as they are buried into the sediment and deployed cultch would not add appreciable weight per unit area.

No Action

There would be no changes to utilities or public services.

4.3.11 Water Resources

Affected Environment

Mississippi nearshore water resources consist of wetlands, shorelines, bays, intertidal and subtidal areas, and open water habitat.

Environmental Consequences

Deployment of cultch material could cause temporary increases in turbidity and suspended sediment concentrations in the water column. However, this effect would be minor, and localized expected to last a few hours until particles have settled out. Certification of the project by the Mississippi Department of Marine Resources has been issued in compliance with the Mississippi Coastal Wetlands Protection Act. Miss. Code Ann. § 49-27-1, *et seq*. This certification also serves as the coastal zone consistency certification for the purposes of the Coastal Zone Management Act in accordance with the Mississippi Coastal Program (DMR-090383; March 6, 2009). Coastal Zone consistency certification has been issued by the MDMR for the Phase I ERP/EA. A Nationwide Permit 48 for shellfish aquaculture has been issued by the U.S. Army Corps of Engineers for oyster cultch placement on existing reefs in the Mississippi Sound (SAM-2007-00316-MFM; April, 11, 2011). Best management practices would be implemented in accordance with applicable permit conditions. Permitted and potential cultch placement areas are identified on Figure 10.

No Action

There would be no changes to water resources.

4.3.12 Cumulative Impacts

The Mississippi Coastal Improvement Plan (MsCIP) (USACE, 2009) is a key planning document which addresses coastal restoration and protection for the Mississippi Sound. While the Mississippi Oyster Cultch Restoration project is not a part of the MsCIP plan, the project is consistent with the goals of this regional plan as it restores and enhances coastal habitats and ecosystems in coastal Mississippi. Deployment of oyster cultch materials would occur within designated shellfish harvesting zones of the western Mississippi Sound and would restore and enhance existing oyster cultch areas. Thus, there would be no conflicting uses for the substrate covered by these deployments. The project would provide additional oyster production in the western Mississippi Sound and it would increase commercial oyster harvesting opportunities. Although the Mississippi oyster cultch restoration proposed early restoration project has potential short-term negative effects, on balance, the proposed project has positive effects that are consistent with long-term planning goals, and contribute beneficially to the Mississippi Sound environment. Additionally, all immediate effects are relatively local and geographically disparate.

No Action

There would be no cumulative impacts under the No Action alternative.

4.3.13 Summary

The Trustees believe that the proposed project would enhance oyster production within the Mississippi Sound.

4.4   Mississippi Artificial Reef Habitat

Purpose and Need

Mississippi's nearshore reefs and shallow-water resources were exposed to oil, dispersant, as well as response activities undertaken to prevent, minimize, or remediate oiling from the Spill. Under OPA, the Trustees act on behalf of the public to restore, rehabilitate, replace, or acquire the equivalent of natural resources injured and associated service losses as a result of the Spill. Under the Framework Agreement, the Trustees have the opportunity to negotiate with BP to fund early restoration projects. The purpose of a Mississippi artificial reef habitat restoration project implemented under OPA and the Framework Agreement is to begin to restore, rehabilitate, replace or acquire the equivalent of Mississippi's secondary production of invertebrate infaunal and epifaunal biomass.

General Project Information

Artificial reefs are located in offshore and nearshore waters of the state of Mississippi. Offshore reefs provide habitat for larval and juvenile recruitment, survival, growth and reproduction for a variety of important species that are currently under the Federal Reef Fish Management Plan

Nearshore artificial reefs provide valuable hardbottom habitat with foraging and shelter sites for various species of larvae and sessile epifauna and infauna (invertebrates and vertebrates). There are 67 existing nearshore artificial reefs in Mississippi waters which are managed by MDMR's

Artificial Reef Bureau. The project consists of the restoration and enhancement of these existing reefs that are approximately 3 acres in size (201 acres in total) using crushed limestone. This material would be placed within existing artificial reef habitat footprints to enhance approximately half of the area (100.5 acres) resulting in reefs with a 4-6 inch profile. For project details, please refer to Section 3.2.3.2.

4.4.1 Aesthetics and Visual Resources

Affected Environment

The proposed project area consists of open water viewsheds within nearshore areas of the Mississippi Sound.

Environmental Consequences

The use of barges and large equipment could have a temporary visual impact during the time of project implementation. The deployment time would be short and therefore any visual impacts would be short as well. The artificial reef profile is low (4-6 inches high) but may extend above the water surface during low tides. However, it is expected that the deployed natural limestone would blend well with the surrounding substrate, thereby not adversely affecting aesthetic and visual resources.

No Action

Under the No Action alternative, aesthetics and visual resources would not be impacted.

4.4.2 Air Quality

Affected Environment

The air quality within coastal Mississippi is in attainment with the National Ambient Air Quality Standards (MDEQ, 2010).

Environmental Consequences

Project implementation would require the use of heavy equipment which could temporarily lead to air pollution due to equipment exhaust. Fine particulate matter associated with the crushed limestone may become airborne during the deployment process. Available best management practices would be employed to prevent, mitigate, and control potential air pollutants during project implementation. Any minor air quality impacts would be localized and short in duration.

No Action

There would be no changes in air quality.

4.4.3 Biological Resources

Affected Environment

The nearshore biological resources of Mississippi consist of a diverse group of marine species and ecologically valuable habitats.

Although coastal Mississippi harbors a number of federally-listed threatened, endangered, or candidate species not all of these typically occur in the nearshore habitat of the project area. Table B-1 lists the federal and state listed threatened and endangered species that potentially could occur in the project area. The green sea turtle, hawksbill sea turtle, Kemp's ridley sea turtle, leatherback sea turtle, loggerhead sea turtle, and West Indian manatee do not have more than a transient occurrence, if any, within the proposed project area. The listed least tern and piping plover use beach, mudflat, and riverine habitats not the nearshore habitat of the project area. In addition, table B-1 lists several federally listed whale and coral species that do not occur in the project area. A discussion of Gulf sturgeon occurrence and EFH compliance is presented in the Environmental Consequences section below.

Environmental Consequences

Short-term disturbances to the water column and benthic organisms could occur when the project is implemented. The deployed material is expected to displace or cover some infauna and epifauna. However, many epifaunal organisms are mobile and would be minimally affected by the settling material. Biological impacts would be temporary. Overall the project would result in an improved marine ecosystem especially for sessile organisms and fish species of commercial and recreational value. Nearshore artificial reefs would provide valuable hardbottom habitat with foraging and shelter sites for various species of larvae and sessile epifauna and infauna (invertebrates and vertebrates). MDMR issues certificates of waivers under the Mississippi Coastal Wetlands Protection Act for work on nearshore artificial reef projects. All effort would be made to avoid existing environmentally sensitive areas such as oyster reefs, emergent and submerged aquatic vegetation, and other live bottom communities. The FWS evaluated whether this project would affect the West Indian manatee has concluded that the project may affect, but is not likely to adversely affect the West Indian manatee.

Within the Magnuson-Stevens Fishery Conservation and Management Act, essential fish habitat (EFH) is defined as types of waterbodies, habitats, and substrates necessary for federally and regional fishery management council managed fish to complete various life history stages such as breeding, spawning, feeding or growth and survival to maturity. Within the proposed project areas, habitat that falls within this designation includes the water column and both hard and soft substrates such as silt, clay, sand, rock, and shell. The threatened Gulf sturgeon (*Acipenser oxyrinchus desotoi*) is an anadromous fish that overwinters in the Gulf of Mexico and adjacent estuaries and bays.

Deployment of artificial reef material is likely in the spring and fall. Ross et al. (2008) performed telemetry studies which indicated that Gulf sturgeon were present in Mississippi Sound habitats from October through March although primarily November through March. Therefore, sturgeon would not be present in the proposed project area during time of deployment. In addition, these telemetry studies showed that once Gulf sturgeon leave the freshwater riverine spawning habitats they typically are found in the shallow water habitats of the barrier island passes with no occurrences in the nearshore habitats of the proposed project. This suggests that sturgeon presence in the nearshore environments is minimal, sporadic, and only occurs during seasonal migrations to barrier island shallow waters. Lastly, the foraging habitat of sturgeon is mainly soft, sandy substrate not the hard substrate of existing artificial reef. Although the proposed project would only deploy materials on existing reef footprints, a limited amount of soft substrate, and sturgeon foraging habitat, could potentially be lost during and following deployment. Therefore, due to the life cycle of the Gulf sturgeon, its preferred foraging habitat, and the location and timing of material placement it is likely that the Gulf sturgeon would not be impacted or would only be minimally impacted by the proposed project. The National Marine Fisheries Service has evaluated whether the project may affect the Gulf Sturgeon and it critical habitat under Section 7 of the Endangered Species Act, and has concluded the project is not likely to adversely affect the Gulf sturgeon or its critical habitat. The National Marine Fisheries Service has evaluated whether the project may affect the Leatherback sea turtle, the Loggerhead sea turtle, the Kemp's Ridley sea turtle, the Green sea turtle, and the Hawksbill sea turtle has concluded that the project is not likely to adversely affect those species.

4.4.4   Cultural Resources

Affected Environment

The area of potential effect (APE) used during reviews under Section 106 of the National Historic Preservation Act includes the areas of direct and indirect impact. For this proposed project it consists of the footprint of the artificial reef material placement. Shipwrecks, their associated artifacts and other cultural resources could potentially be affected in the project area.

Environmental Consequences

National Historic Preservation Act Section 106 was considered during the USACE and MDMR environmental permitting process. No shipwrecks or other cultural resources are known to exist in the project area. Consultation with the Mississippi Department of Archives and History (MDAH) was initiated to determine the presence or absence of historic, archaeological, or culturally significant sites. The State Historic Preservation Officer (SHPO) determined that no cultural resources were likely to be affected during implementation of the proposed project (MDAH Project Log #09-174-11, October 11, 2011). If any culturally or historically important resources are identified during project preparations, such sites would be avoided during site selection. A complete review of this project under Section 106 of the National Historic Preservation Act will be completed prior to project implementation.

No Action

Cultural resources would not be impacted.

4.4.5   Geology, Soils, and Sediments

Affected Environment

The targeted nearshore deployment would be implemented within existing nearshore artificial reefs footprints which consist of hard reef substrate of limestone or concrete as well as a very limited amount of soft sediments of sand, silt, or clay.

Environmental Consequences

There should be minimal impacts to geology, soils, or sediments. Artificial reef material would only be placed on existing reef footprints. This action would mainly cover existing hard substrates although it could potentially replace a limited amount of soft sedimentary substrates with hard substrates. The project would create low profile alterations on average 4 inches (although no more than 6 inches) above the substrate to localized areas of the seafloor. The placed limestone would assist in stabilizing the coastline during storm events and reduce the mobilization of sediment.

No Action

There would be no changes to existing geology, soils, and sediment.

4.4.6   Land Use

Affected Environment

The proposed project areas consist of open water within the Mississippi Sound and do not include terrestrial or shoreline areas.

Environmental Consequences

Implementation of the project would not disrupt existing land uses, shoreline areas, or wetlands. However, the project would be consistent with the coastal wetlands use designations set forth in the Mississippi Coastal Program and any other applicable local zoning requirements.

No Action

There would be no impact to existing land uses.

4.4.7   Noise

Affected Environment

The current noise levels are minimal on the open water of the proposed project areas.

Environmental Consequences

The project requires the use of heavy equipment, boats, and barges for implementation which could emit noise. Wildlife and humans in the area could be impacted. Noise levels above current background noise levels would be limited to the short duration of project deployment.

No Action

If the project were not implemented, there would be no changes in current noise levels.

4.4.8   Socioeconomics and Environmental Justice

Affected Environment

In 2009, the seafood industry in the State of Mississippi provided approximately 6,400 jobs and generated $289 million in sales and $113 million in personal income (NOAA Fisheries 2011a). The recreational fishing industry provided approximately 3,200 jobs and generated $417 million in sales and $106 million in personal income (NOAA Fisheries 2011a). The commercial fishing industry provided approximately 1,200 jobs and generated $61 million in sales and $19 million in personal income (NOAA Fisheries 2011a). The shellfish fishing sector provided approximately 1,100 jobs (Posadas and Posadas, 2011).

Environmental Consequences

There should be no adverse social, economic, health, or environmental impacts to local communities due to this project. Development of 100 acres of nearshore artificial reef would enhance the existing MDMR artificial reef management efforts. In fact, improved marine habitat would provide greater economic and commercial resources for local citizens and local businesses.

The project would not have a disproportionate effect on any particular group of people or individuals, including low income or minority populations.

No Action

There would be no socioeconomic impacts or environmental justice considerations if the project were not constructed.

4.4.9 Public Access/Recreation

Affected Environment

Productivity within placed nearshore artificial reefs develops within the first year. Access to the nearshore artificial reef areas will remain available.

Environmental Consequences

During placement of artificial reef material, public access to the deployment area would be temporarily restricted. However, the deployment time in any given area is very short and therefore impacts to public access in the area are considered minimal.

No Action

There would be no changes to public access.

4.4.10 Utilities and Public Services

Affected Environment

Potential utilities or public services within the underwater area of the project are expected to be buried beneath the sediment, unless storms have exposed utilities that were buried in the past.

Environmental Consequences

Deployment of artificial reef material would cover a targeted area. It is anticipated that the proposed project would not adversely impact any buried utilities or public services in the proposed project area. Areas of known or suspected exposed utilities, if any, would be avoided for limestone placement.

No Action

There would be no change to utilities or public services.

4.4.11 Water Resources

Affected Environment

Mississippi's water resources consist of nearshore coastal wetlands, shorelines, bays, intertidal and subtidal areas, and open water habitat.

Environmental Consequences

Deployment of artificial reef material would cause slight disturbances to the sea floor sediment which could temporarily (hours) increase the turbidity and suspended sediment concentration in the water column. Deployment would occur in areas where resuspension of sediment and hence increased turbidity occurs during storms. Best management practices would be used when implementing the project to minimize turbidity increases. Certification of the project by the MDMR has been issued in compliance with the Mississippi Coastal Wetlands Protection Act. Miss. Code Ann. § 49-27-1, *et seq*. This certification also serves as the coastal zone consistency certification for the purposes of the Coastal Zone Management Act and the Mississippi Coastal Program (DMR-120097; October 28, 2011). Coastal zone consistency certification has been issued by the MDMR for the Phase I ERP/EA. A Nationwide Permit 4 has been issued for material deployment on existing reefs in the Mississippi Sound (SAM-2011-01777-SPG, November 30, 2011). The permit includes developed reef areas as well as undeveloped acreage within 67 existing sites. All conditions within the permit would be adhered to.

No Action

There would be no changes in water resources.

4.4.12 Cumulative Impacts

The Mississippi Coastal Improvement Plan (MsCIP) (USACE, 2009) is a key planning document which addresses coastal restoration and protection for the Mississippi Sound. While the Mississippi Artificial Reef Habitat project is not a part of the MsCIP plan, the project is consistent with the goals of this regional plan as it restores and enhances coastal habitats and ecosystems in coastal Mississippi. Deployment of crushed concrete or limestone would occur within designated nearshore artificial reef areas. Thus, there would be no conflicting uses for the substrate covered by these deployments. The project would also have ecological benefits for the nearshore area of the Mississippi Sound. Although the Mississippi Artificial Reef proposed early restoration project has potential short-term negative effects, on balance, the proposed project has positive effects that are consistent with long-term planning goals, and contribute beneficially to the Mississippi Sound environment. Additionally, all effects are relatively local and geographically disparate.

No Action

There would be no cumulative impacts resulting from a no action alternative.

4.4.13 Summary

Overall, this project would enhance the Mississippi coastal and marine ecosystem. The beneficial ecological impacts are expected to far outweigh any short-term, adverse impacts from deployment of artificial reef material. The Trustees believe that the proposed project would increase Mississippi's secondary production of invertebrate infaunal and epifaunal biomass.

4.5   Marsh Island (Portersville Bay) Marsh Creation

Purpose and Need

Marshes in Alabama were exposed to oil, dispersant, and response activities undertaken to prevent, minimize, or remediate oiling from the Spill. Under OPA, the Trustees act on behalf of the public to restore, rehabilitate, replace, or acquire the equivalent of natural resources injured and associated service losses as a result of the Spill. Under the Framework Agreement, the Trustees have the opportunity to negotiate with BP to fund early restoration projects. The purpose of an Alabama marsh habitat restoration project implemented under OPA and the Framework Agreement is to begin to restore, rehabilitate, replace or acquire the equivalent of Alabama's marsh resources.

General Project Information

The proposed NRDA early restoration project Marsh Island (Portersville Bay) Marsh Creation Project involves the creation of salt marsh along Marsh Island, a state-owned island in the Portersville Bay portion of Mississippi Sound, Alabama. This project would add approximately 50 acres of salt marsh to the existing 24 acres of Marsh Island, through the construction of a permeable segmented breakwater, the placement of sediments and the planting of native marsh vegetation. Additionally, this project would protect the existing salt marshes of Marsh Island, which have been experiencing significant losses due to chronic erosion. For more project details, please refer to Section 3.2.4.

The environmental assessment for this project is based on general information regarding the proposed design and construction of the project currently available at this time. Because the information needed to finalize an analysis under NEPA is not available, this project would be subject to further environmental analysis and public review once sufficient information is developed to provide for that analysis. A general project footprint was used as a basis to make conservative assumptions that were used to evaluate a range of possible impacts. Any dimensions or description of site features are approximate, based on a typical conceptual design that meets the purpose and need for the project. Specific information on construction methods and design details will be developed at a later date. During the design process and borrow area siting, mitigation measures (e.g., conservation design standards, erosion and sedimentation best management practices, project timing) would be implemented to minimize impacts to the environment.

4.5.1 Aesthetics and Visual Resources

Affected Environment

The proposed project area consists of open water and marshland.

Environmental Consequences

The proposed project would involve the placement of a permeable segmented breakwater constructed of riprap, wave attenuation devices or other similar materials. Additionally, hazard to navigation signage would be placed along the breakwaters. During construction, dredges, marsh buggies, barges, small tugs and other machinery would be on-site. During construction, impacts to aesthetic and visual resources due to machinery and construction activities would be short-term and temporary. Once construction is completed, the permeable segmented breakwater and hazard to navigation signage would remain in place. While such man-made objects are not normally found in this location, they are common sites all along the coast. Therefore, the benefits of project construction greatly out-weigh any impacts to aesthetic and visual resources.

No Action

Aesthetics and visual resources would not be impacted if the project was not implemented. If the project is not implemented further erosion and ultimate loss of the existing marshes and accompanying habitat would occur.

4.5.2   Air Quality

Affected Environment

The air quality within coastal Alabama is in attainment with the National Ambient Air Quality Standards (NAAQS) (USACE, 2009).

Environmental Consequences

Short term, minor, temporary impacts to local air quality may result from vehicle operation during construction. Project implementation would require the use of heavy equipment which could temporarily lead to air pollution due to equipment exhaust. However, no air quality permits are required for this type of project and no violations of state air quality standards are expected from a project of this type and scope. Any available best management practices would be employed to prevent, mitigate, and control potential air pollutants during project implementation. Any minor air pollution that does occur would be localized and short in duration.

No Action

If the project is not implemented, no changes in air quality would occur.

4.5.3   Biological Resources

Affected Environment

West Indian manatees may occasionally occur in Mississippi Sound. Gulf Sturgeon have been known to occur in Mississippi Sound. However the project area is not designated as critical habitat for the Gulf Sturgeon. Leatherback, Kemp's ridley, Hawksbill, Green and Loggerhead sea turtles all may occur in the project area.

Mississippi Diamond-backed terrapins are a species of special concern and are known to exist in the project area. Any possible impacts to Mississippi diamond-backed terrapins, if they occur at all, are expected to be minor and temporary. Therefore, no significant or long-term adverse impacts are expected.

Existing salt marsh on Marsh Island may be temporarily impacted by construction activities, such as marsh buggy operations, gathering of marsh plant plugs, and other similar activities. However, these impacts would be temporary. Additionally, the selected construction contractor would be required to correct any adverse impacts to existing wetlands. Further, the construction of the proposed breakwater would protect the existing marsh, abating long term erosion at the site. CWA Section 10/404 permits and Water Quality Certification from the Mobile District of the U.S. Army Corps of Engineers (Corps) and the Alabama Department of Environmental Management (ADEM) would be required.

Based on submerged aquatic vegetation (SAV) surveys conducted in 2002, 2008 and 2009 by the Alabama Department of Conservation and Natural Resources, there are no known SAVs in the project area. However, an SAV survey would be conducted as part of the environmental investigations conducted as part of the design, engineering and permitting phase of the project.

Environmental Consequences

The National Marine Fisheries Service evaluated the proposed project and concluded that the project would not adversely affect EFH, and overall, would likely benefit federally managed fishery species. Additionally, the National Marine Fisheries Service evaluated the project under Section 7 of the Endangered Species Act and determined that it will not adversely affect the Gulf sturgeon, the Leatherback sea turtle, the Kemp's ridley sea turtle, the Hawksbill sea turtle, the Green sea turtle, and the Loggerhead sea turtle. No impacts to threatened and/or endangered species are expected. Should dredging activities be implemented during the summer months an observer would be watching for manatees to ensure that collisions would be avoided.

Impacts to Species of Special Concern would be temporary and short term. Project construction would result in increased Mississippi Diamond-backed Terrapin foraging and nesting habitat.

Any impacts to the existing salt marsh would be temporary and/or repaired upon project completion.

No impacts to submerged aquatic vegetation are anticipated.

CWA Section 10/404 permits and Water Quality Certification from the Mobile District of the U.S. Army Corps of Engineers (Corps) and the Alabama Department of Environmental Management (ADEM) would be required and obtained. It is too early in the design and engineering phase of the project to obtain those permits now.

The Alabama Department of Environmental Management (ADEM) reviewed the project proposal under the Coastal Zone Management Act and determined that the proposal is consistent with the enforceable policies of the Alabama Coastal Management Program to the extent that these activities have been defined by the current level of planning and design in the Phase I ERP/EA.

No Action

No action to limit environmental consequences would result in a slower recovery of the affected salt marshes. Non-implementation of the proposed project would result in further erosion and ultimate loss of the existing marshes.

4.5.4   Cultural Resources

Affected Environment

This proposed project has the potential to affect cultural resources if such resources are present. A search for known cultural resources in the project area would be completed as required by USACE permit conditions.

Environmental Consequences

The Trustees would comply with Section 106 of the National Historic Preservation Act, as part of the site investigations for the design and engineering process, to avoid or mitigate any potential effects to cultural resources that are located within the project area.

No Action

No project implementation would have the potential for adverse impacts to any existing cultural resources in the existing marsh from accelerated erosion that could occur if the project is not implemented.

4.5.5   Geology, Soils, and Sediments

Affected Environment

Geotechnical investigations of possible sediment borrow sites would be conducted. This would include an analysis of possible impacts of removing sediments from the borrow site. Sediments and soils along the existing marsh would be stabilized by the construction of the permeable segmented breakwater.

Environmental Consequences

No substantial adverse effects to sediment quality, soil, or geologic conditions would be expected as a result of the project.

No Action

No project implementation would result in further erosion and loss of sediment from the existing marsh.

4.5.6   Land Use

Affected Environment

The current land use of the project site is conservation and preservation. No change in this status would take place.

Environmental Consequences

No changes in land use or land use patterns would result.

No Action

No project implementation would result in further erosion and ultimate loss of the existing marshes.

4.5.7 Noise

Affected Environment

Short term, minor, temporary noise impacts from marsh buggy, dredge and other machinery operation during construction is expected.

Environmental Consequences

Machinery and equipment used during construction would generate noise. This noise may disturb wildlife and humans using the area. However, once built, the proposed project would not cause appreciable noise impacts.

No Action

No project implementation would result in further erosion and ultimate loss of the existing marshes.

4.5.8 Socioeconomics and Environmental Justice

Affected Environment

The proposed project site is located in an area of wildlife habitat (or open water), and no housing would be affected because none exists in the proposed project site. Bayou la Batre and Coden are the closest communities to the project site.

Environmental Consequences

The proposed restoration project would have no adverse social or economic impacts on neighborhoods or communities. The project could result in minor short-term beneficial impacts on the local economy due to temporary employment or local spending during project construction. The proposed project would not have any adverse effect on human or socioeconomic resources; therefore, the proposed project complies with the requirements of Executive Order 12898.

No Action

No action would result in further erosion and ultimate loss of the existing marshes.

4.5.9 Public Access/Recreation

Affected Environment

The waters and shorelines along Marsh Island and in the vicinity of the project site are utilized for fishing, boating, waterfowl hunting and other recreational uses.

Environmental Consequences

Public access and recreational use may temporarily be affected during construction activities. Because implementation time for the proposed project would be relatively short, the impact would be short in duration.

No Action

No action would result in further erosion and ultimate loss of the existing marshes, resulting in the loss of recreational use of the existing marsh and shorelines.

4.5.10 Affected Environment

There are no public utilities and/or services associated with the project site.

Environmental Consequences

No impacts are expected from implementation of this project.

No Action

No action would result in further erosion and ultimate loss of the existing marshes.

4.5.11 Water Resources

Affected Environment

The project area consists of marshland and open water.

Environmental Consequences

Dredging of sediments from the borrow site, the placement of sediments for marsh creation and the construction of the permeable segmented breakwater may result in short term, minor, temporary impacts to water quality, specifically short term elevations in turbidity. Best management practices along with other avoidance and mitigation measures required by state and federal regulatory agencies would be employed to minimize any water quality and sedimentation impacts. Section 10/404 and Water Quality Certifications would be required and all permit conditions would be adhered to.

No Action

No action would result in further erosion and ultimate loss of the existing marshes.

4.5.12 Cumulative Impacts

Cumulative impacts for this project will be addressed as part of future environmental analyses under NEPA.

4.5.13 Summary

At this time, sufficient information is not available to determine whether or not this project would have a significant impact on the human environment. A complete NEPA analysis will be completed for this project once sufficient information regarding the project design becomes available.

4.6   Alabama Dune Restoration Cooperative Project

Purpose and Need

Department of the Interior and Alabama dunes were exposed to oil and dispersants and/or affected by response activities undertaken to prevent, minimize, or remediate oiling from the Spill. Under OPA, the Trustees act on behalf of the public to restore, rehabilitate, replace, or acquire the equivalent of natural resources injured and associated service losses as a result of the Spill. Under the Framework Agreement, the Trustees have the opportunity to negotiate with BP to fund early restoration projects. The purpose of an Alabama dune restoration project implemented under OPA and the Framework Agreement is to begin to restore, rehabilitate, replace or acquire the equivalent of Department of the Interior and Alabama dune resources.

General Project Information

This proposed NRDA early restoration project would provide early restoration for dune habitat and beach mice injured as a result of the Spill. Dune vegetation in the Bon Secour NWR, BLM Fort Morgan properties, and coastal areas in Alabama has been injured by exposure to *Deepwater Horizon* oil and/or response activities. The project is needed to help restore an area of beach where oiling and the extensive use of all-terrain vehicles and heavy equipment has inhibited plant growth and prevented the natural seaward expansion of the dunes since May 2010.

This project involves planting native vegetation and installing sand fencing and signage. No new access roads or staging areas would be built as part of this project. Vehicles would use existing roads and parking areas. All participants involved in the project would follow rules established to minimize noise, foot traffic and human presence across ecologically sensitive areas. The planting portion of the project would occur during the growing season (approximately March-June). Sand fence installation could be completed at any time during the year, and would be installed when nesting sea turtles would not be impacted. Sand fencing would be installed as per Alabama Department of Environmental Management Coastal Sand Fencing Construction Guidelines (Appendix C). For project details, please see Section 3.2.5.

NEPA compliance

NEPA requires Federal agencies to analyze their proposed actions to determine if they could have significant environmental effects. Over time, through study and experience, agencies may identify activities that do not need to undergo detailed environmental analysis in an environmental assessment (EA) or an environmental impact statement (EIS) because the activities do not individually or cumulatively have a significant effect on the human environment. Agencies can define categories of such activities, called categorical exclusions (CXs), in their NEPA implementing procedures, as a way to reduce unnecessary paperwork and delay.

If an agency determines that a proposed activity fits within the description of one or more categorical exclusions and that there are no extraordinary circumstances that might cause significant environmental effects, no additional NEPA review is required and the agency can proceed with the activity without preparing an EA or EIS. Categorical exclusions are an essential tool in facilitating NEPA implementation and concentrating environmental reviews on instances of potential impacts. A CX is a form of NEPA compliance, without the analysis that occurs in an EA or an EIS. Categorical exclusions are not exemptions or waivers of NEPA review; they are simply one type of NEPA review (CEQ issued NEPA Guidance on Categorical Exclusions on November 23, 2010.)

The U.S. Fish and Wildlife Service's NEPA Procedures in Departmental Manual 516 DM 2.3A (3) and 516 DM 2, Appendix 2, requires that before a CX is used the list of "extraordinary circumstances" be reviewed for applicability. When no "extraordinary circumstances" exist, neither an EA nor an EIS is required (40 CFR 1508.4). Extraordinary circumstances are factors or circumstances in which a normally excluded action may have a significant environmental effect that then requires further analysis in an environmental assessment or environmental impact statement (CEQ Memorandum 2010).

After undergoing NEPA review, the Trustees determined this project would meet two resource management categorical exclusions as described in 516 DM6 Appendix 1, Section 1.4, nos. 3 and 11, and Bureau of Land Management Department Manual 516 DM 11.9 These categorical exclusions are:

(3) The construction of new, or the addition of, small structures or improvements, including structures and improvements for restoration of wetland, riparian, instream, or native habitats, which result in no or only minor changes in the use of the affected local area. The following are examples of activities that may be included.
i. The installation of fences.
ii. The construction of small water control structures.
iii. The planting of seeds or seedlings and other minor revegetation actions.
iv. The construction of small berms or dikes.
v. The development of limited access for routine maintenance and management purposes.

(11) Natural resource damage assessment restoration plans, prepared under sections 107, 111, and 122(j) of the Comprehensive Environmental Response Compensation and Liability Act (CERCLA); section 311(f)(4) of the Clean Water Act; and the Oil Pollution Act; when only minor or negligible change in the use of the affected areas is planned.

Because the dune restoration project involves planting and other minor revegetation actions, installing sand fencing and signage, and would result in only minor or negligible change in the use of the project area, U.S. Fish and Wildlife Service determined that it would apply categorical exclusions i and iii described above to this project.

A NEPA Compliance Checklist (FWS Form 3-2185) and Environmental Action Statement (EAS) were prepared to document the use of the categorical exclusions. An EAS is "a Service-required document prepared to improve the Service's administrative record for categorically excluded actions that may be controversial, emergency actions under CEQ's NEPA regulations (40 CFR 1506.1 1), decisions based on EAs to prepare an EIS, and any decision where improved documentation of the administrative record is desirable, and to facilitate internal program review and final approval when a FONSI is to be signed at the FWS-WO and FWS-RO level (550 FW3)." The NEPA Compliance Checklist was used to help determine the applicability of a CX.

Since project scopes, environmental conditions and regulatory requirements can change over time, the use of these CXs would be reviewed for their continued applicability to the project before implementation.

Summary

No threatened or endangered species, or eligible cultural sites or historic properties would be affected as a result of implementing this project. The FWS evaluated this project under Section 7 of the Endangered Species Act and has concluded that the project would not have an adverse impact to the endangered Alabama beach mouse and its critical habitat, the Loggerhead sea turtle, the Kemp's Ridley sea turtle, the piping plover and its critical habitat, and the snowy plover. The project will be reviewed under Section 106 of the National Historic Preservation Act prior to project implementation.

The Alabama Department of Environmental Management (ADEM) reviewed the Alabama Dune Restoration Cooperative Project proposal under the Coastal Zone Management Act and determined that the proposal is consistent with the enforceable policies of the Alabama Coastal Management Program to the extent that these activities have been defined by the current level of planning and design in the Phase I ERP/EA.

The National Marine Fisheries Service has evaluated the proposed Alabama Dune Restoration Cooperative Project proposal and concluded that the project would not adversely affect EFH, and overall, would likely benefit federally managed fishery species.

Overall, this project would enhance the Alabama dune ecosystem on the Bon Secour NWR, BLM Fort Morgan properties, Gulf Shores State Park and within the City of Gulf Shores, and City of Orange Beach. The Trustees determined that the proposed activity qualifies for two categorical exclusion(s) and that there are no extraordinary circumstances that might cause significant environmental effects. Further, the Trustees believe the project would have no potential adverse impact on the quality of the human environment, either individually or cumulatively. Accordingly, no additional NEPA analysis for this project is required at this time.

4.7   Florida Boat Ramp Enhancement and Construction Project

Purpose and Need

In the Florida panhandle, boaters were deterred from using public boat ramps by the Spill. Under OPA, the Trustees act on behalf of the public to restore, rehabilitate, replace, or acquire the equivalent of natural resources injured and associated service losses resulting from the Spill.

Under the Framework Agreement, the Trustees have the opportunity to negotiate with BP to fund early restoration projects. The purpose of a restoration project to improve boating opportunities implemented under OPA and the Framework Agreement is to begin to restore, rehabilitate, replace or acquire the equivalent of recreational service losses in Florida attributable to the Spill.

General Project Information

This proposed NRDA early restoration project would provide early restoration for lost human use services of natural resources injured as a result of the Spill. In the Florida panhandle, boaters were deterred from using public boat ramps during the Spill because pollutants in the water made taking boating trips less desirable. Furthermore, a number of boat ramps were utilized by response equipment and personnel, preventing the public from accessing boat ramps for recreational use. Navy Point and Galvez Landing boat ramps, among numerous others, were used as staging areas from May to July 2010 to facilitate vessels of opportunity deploying boom and engaging in other response activities. This project would help restore the impacts to recreational activities (e.g., boating and fishing) in Florida attributable to the Spill. The two new ramps proposed and the enhancement at the existing Galvez Landing and Navy Point boat ramps are expected to reduce boat traffic congestion at other ramps in the area.

The Pensacola Bay system is located in northwest Florida in Escambia and Santa Rosa Counties. The Pensacola Bay watershed includes three major river systems: the Escambia, Blackwater, and Yellow rivers. The major rivers discharge into an estuarine system that includes Escambia Bay, Pensacola Bay, Blackwater Bay, East Bay, and Santa Rosa Sound, which discharge into the Gulf of Mexico. The watershed covers nearly 7,000 square miles, about one-third of which is in Florida (Thorpe et al., 1997). The Perdido River is located in Baldwin County, Alabama and Escambia County, Florida, with the state line bisecting the river and bay. The Perdido River discharges into Perdido Bay about 15 miles west of Pensacola, Florida. Both Perdido and Pensacola bays have an average depth of about 3 meters, with a salinity range of 0-32 ppt. Both bay systems are composed of riverine and estuarine habitat types, each with an abundance of natural resources.

This project would build two new boat ramps and enhance two existing boat ramps, providing boaters with enhanced access to public waterways within Pensacola Bay, Perdido Bay, and offshore areas. The Navy Point boat ramp is an existing ramp in Pensacola Bay, in a developed, residential area. The Galvez Landing boat ramp is an existing ramp in Perdido Bay, in a residential area. The Mahogany Mill boat ramp, in Pensacola Bay, is proposed to be built in a commercial and industrial area. The Perdido River boat ramp is proposed to be built in a less developed area than the other three. There are no parks or wildlife refuges near the project sites. For project details, please see Section 3.2.6.

Mahogany Mill – The construction of this new boat ramp would require 496 cubic yards of sediment to be dredged, and would impact .02 acres of wetlands (Consolidated Wetland Resource Field Permit and Sovereign Submerged Lands Authorization, Florida DEP, July 12, 2010). This site is located on Bayou Chico which has been the site of various industrial and marine activities going back many years. In the past the water quality was severely impacted by these uses along with urban development and associated runoff. In the last 20 years there have been numerous cleanup projects undertaken at the local, state and Federal levels to improve the Bayou's water quality. Various degrees of success have been and continue to be obtained through these efforts such as industrial site closings, spill protection and prevention plans being implemented, environmental enhancements at commercial ship building and marina facilities and storm water retention and treatment facilities being constructed. The proposed site of the Mahogany Mill facility was a former mahogany wood mill (industrial) that received, milled and treated mahogany wood products. This facility has been closed for over 30 years. A Phase 1 and Phase 2 environmental site assessment was performed on the site prior to the County purchasing the property. Current surrounding property uses are commercial marinas to the north and south and an apartment complex landward on the west side of the access right-of-way to the site. The Florida DEP has evaluated the project construction proposal and granted regulatory and proprietary authorization for the construction of the boat ramp, which includes a Certification of Compliance with State Water Quality Standards (Section 401, PL 92-500). The Army Corps of Engineers has issued an individual permit under Section 10 of the Rivers and Harbors Act and Section 404 of the Clean Water Act for the construction of this boat ramp.

Perdido River – This project consists of constructing a new boat ramp in an area of single family dwellings and apartments. The site is the location of a former single family home site with a covered boat slip in poor condition. Escambia County demolished the original dwelling unit due to its unsafe, dilapidated condition. The adjacent riverside properties include a vacant single family home site (trailer) and an occupied single family home. Most of the nearby properties along the Florida side of the river are developed and used as single family homes; several of these sites have private boat docking facilities. The land across (Alabama) from the proposed site is vacant wooded land. Landward (south) of the site is Hwy 90, a major east-west transportation corridor between Florida and Alabama.

Galvez Landing – This project consists of removing and replacing three piers on an existing boat ramp. The Florida DEP has evaluated the project proposal and has granted regulatory and proprietary authorization (Consolidated Environmental Resource Field Permit and Sovereign Submerged Lands Authorization, Florida DEP, March 17, 2011). This authorization also constitutes a Certification of Compliance with State Water Quality Standards under Section 401 of the Clean Water Act. The Army Corps of Engineers has evaluated the project proposal and has issued a Letter of Permission to begin construction, which serves as a permit authorization under Section 10 of the Rivers and Harbors Act.

Navy Point – This project is consists of 260 cubic yards of maintenance dredging at an existing boat ramp. The Florida DEP has evaluated this project proposal and determined that it will have only minimal or insignificant individual or cumulative adverse impacts on the water resources due to the type, size, nature, location, use and operation of the project. Consequently, the Florida DEP determined that this project is exempt from regulatory requirements to obtain a Florida DEP permit (Letter of Authorization, Florida Department of Environmental Protection, March 9, 2012). This authorization letter also serves as the sovereign submerged lands authorization. The Army Corps of Engineers has evaluated the project proposal and has issued a Letter of Permission to begin construction, which serves as a permit authorization under Section 10 of the Rivers and Harbors Act.

4.7.1 Aesthetics and Visual Resources

Affected Environment

The four proposed boat ramp sites are in already developed areas, surrounded by single or multifamily residential homes, and industrial or commercial buildings.

Environmental Consequences

Both the Navy Point and Galvez Landing enhancements include upgrading deteriorated old dock structures with new docks, which would enhance the safety and aesthetic value of the sites. The Perdido River site had one single family residential home which was flood damaged and abandoned. The derelict building has been demolished and a park site and boat ramp would be built at this location. Mahogany Mill was an old industrial site no longer in use. The site would be redeveloped to add a monument, a park and a boat ramp. The Mahogany Mill ramp construction includes shoreline stabilization by planting native vegetation along the shoreline at this site.

No Action

Aesthetics and visual resources would not be impacted if the project were not implemented.

4.7.2 Air Quality

Affected Environment

Air quality within the Florida panhandle is in attainment with the National Ambient Air Quality Standards (http://www.epa.gov/airquality/urbanair/sipstatus/reports/fl_areabypoll.html).

Environmental Consequences

Project implementation would require the use of heavy equipment which could temporarily lead to air pollution due to equipment exhaust. Available best management practices would be employed to prevent, mitigate, and control potential air pollutants during project implementation. Any minor pollution that does occur would be localized and short in duration. No air quality related permits would be required. Project implementation could increase boat traffic on the river, which could increase boat exhaust fumes.

No Action

No Action would result in no changes in air quality.

4.7.3 Biological Resources

Affected Environment

Gulf sturgeon, manatees, sawfish and sea turtles (Kemp's Ridley, loggerhead, leatherback, and green) may visit the waters of the four ramp locations. The Navy Point project is located in designated Gulf sturgeon critical habitat. The remaining boat ramp projects are not located in designated critical habitat. Smalltooth sawfish are not likely to be encountered at any of the project sites. Their current distribution has contracted to peninsular Florida and, within that area, they can only be found with regularity off the extreme southern portion of the state (NOAA, National Marine Fisheries Service (NMFS) consultation letter, April 2, 2012).

There are no wading bird rookeries at any of the sites; however wintering piping plovers, least terns and wood storks may occasionally visit the sites. Additional state-listed species may also occur in the area. There are no known bald eagle nests at any of the sites, but due to the heavily wooded area surrounding the Perdido River site, there is potential for nesting in the area. If bald eagles would be found nesting within 660' of the construction area, then activities would need to occur outside of nesting season, or a Bald and Golden Eagle Protection Act permit would be required from the USFWS and Florida's Bald Eagle Management Plan guidelines would need to be followed. Potential take of state-listed species is not anticipated, and would require an appropriate permit.

All four proposed sites are located in developed areas with little or no native vegetation in the project area. See Section 4.7 for more information on the proposed siting locations.

Environmental Consequences

Habitat

The proposed boat ramp locations, whether new construction or existing ramps, are all located in developed areas. Boat ramp construction and operation would cause only minimal alteration and/or damage to habitats. No submerged aquatic vegetation, which is habitat for species such as manatees, sea turtles, fish and invertebrates, was observed at the three sites that have already been permitted (Mahogany Mill, Galvez Landing and Navy Point) and it was determined that fish and wildlife resources would most likely be only minimally impacted. The Florida DEP Wetland and Environmental Resource Field permits, which are required for all of the projects except Navy Point, require Best Management Practices (BMPs) for turbidity and erosion control to be implemented. Navy Point, which only consists of maintenance dredging, has been issued regulatory exemption by Florida DEP; however this exemption also requires implementation of BMPs for turbidity and erosion control. This will help minimize the damage and loss of habitats through the same mitigation measures mentioned under Section 4.7.11, Water Resources. All dredging activities would be done in compliance with Florida DEP permit conditions. These include:

- Measures to prevent spoil material from entering waters of the State,
- Monitoring turbidity at the dredge and spoil disposal sites,
- Take immediate corrective actions if a disposal site leaks or breaks, and
- After recontouring, replanting vegetation of the size, densities and species as is present in the adjacent areas if the area dredged is vegetated (Consolidated Wetland Resource Field Permit and Sovereign Submerged Lands Authorization, Florida DEP, July 12, 2010).

As specified in the Magnuson-Stevens Fishery Conservation and Management Act, Essential Fish Habitat (EFH) consultation is required for federal actions that may affect EFH. The NOAA Restoration Center determined that this project would not adversely affect EFH, and overall would likely benefit federally managed fisheries species (Memorandum, Essential Fish Habitat (EFH) review of the Phase I Early Restoration Projects, February 14, 2012).

The Perdido River and Navy Point sites may attract invasive or nuisance species due to the heavily wooded neighboring areas. The remaining sites are urbanized and pose little risk of attracting nuisance species. Precautions would be taken to prevent soil disturbance which would attract invasive plant species.

Marine Mammals
Escambia County is not listed as one of the 36 Florida coastal and inland counties in which manatees regularly occur (USFWS Biological Opinion, 2011 Manatee Key). Manatees would not be attracted to the area of the permitted boat ramps due to the lack of submerged vegetation for foraging at the sites. The project sites are not adjacent to manatee protection zones so the risk of collision around the boat ramps is low.

The Manatee Key is a tool that has been used by the Army Corps' Regulatory Division since 1992 to assist in making its effect determinations. For certain activities determined to be "may affect, but not likely to adversely affect" using the 2011 Manatee Key, the Service concurs with these determinations and no further consultation with the Service is necessary. These activities include 'all applications for multi-slip facilities proposed to be built in Bay, Dixie, Escambia, Franklin, Gilchrist, Gulf, Hernando, Jefferson, Lafayette, Monroe (south of Craig Key), Nassau, Okaloosa, Okeechobee, Santa Rosa, Suwannee, Taylor, Wakulla and Walton counties" (USFWS Biological Opinion, 2011 Manatee Key).

The Army Corps of Engineers permits discussed above, which are applicable for all the projects, include standard manatee conditions for in-water work. The permittee must comply with the following conditions intended to protect manatees from direct project effects:

- All personnel associated with the project shall be instructed about the presence of manatees and manatee speed zones, and the need to avoid collisions with and injury to manatees. The permittee shall advise all construction personnel that there are civil and criminal penalties for harming, harassing, or killing manatees which are protected under the Marine Mammal Protection Act, the Endangered Species Act, and the Florida Manatee Sanctuary Act.
- All vessels associated with the construction project shall operate at "Idle Speed/No Wake" at all times while in the immediate area and while in water where the draft of the vessel provides less than a four-foot clearance from the bottom. All vessels will follow routes of deep water whenever possible.
- Siltation or turbidity barriers shall be made of material in which manatees cannot become entangled, shall be properly secured, and shall be regularly monitored to avoid manatee entanglement or entrapment. Barriers must not impede manatee movement.
- All on-site project personnel are responsible for observing water-related activities for the presence of manatee(s). All in-water operations, including vessels, must be shut down if a manatee(s) comes within 50 feet of the operation. Activities will not resume until the manatee(s) has moved beyond the 50-foot radius of the project operation, or until 30 minutes elapses if the manatee(s) has not reappeared within 50 feet of the operation. Animals must not be herded away or harassed into leaving.
- Any collision with or injury to a manatee shall be reported immediately to the Florida Fish and Wildlife Conservation Commission (FWC) Hotline at 1-888-404-3922. Collision and/or injury should also be reported to the U.S. Fish and Wildlife Service in

Jacksonville (1-904-731-3336) for north Florida or Vero Beach (1-772-562-3909) for south Florida, and to FWC at ImperiledSpecies@myFWC.com.

- Temporary signs concerning manatees shall be posted prior to and during all in-water project activities. All signs are to be removed by the permittee upon completion of the project. Temporary signs that have already been approved for this use by the FWC must be used. One sign which reads Caution: Boaters must be posted. A second sign measuring at least 8 ½" by 11" explaining the requirements for "Idle Speed/No Wake" and the shutdown of in-water operations must be posted in a location prominently visible to all personnel engaged in water-related activities. These signs can be viewed at MyFWC.com/manatee.

The Florida boat ramp project will adhere to all applicable permit conditions, federal, state and local requirements for the protection of marine mammals during construction.

Sea Turtles, Smalltooth sawfish, Gulf sturgeon
For projects in waters accessible to sea turtles, Smalltooth sawfish, Gulf sturgeon, or Short-nose sturgeon, the permittee must comply with the Army Corps of Engineers' Sea Turtle and Smalltooth sawfish construction conditions (http://www.saj.usace.army.mil/Divisions/Regulatory/sourcebook.htm) and any additional requirements, as appropriate, for the proposed activity (25 July 2011 Memorandum for State Programmatic General Permit (SPGP IV-R1), State of Florida, Department of the Army, Jacksonville District Corps of Engineers). The absence of seagrasses and submerged aquatic vegetation at the proposed sites makes encounters with sea turtles, Smalltooth sawfish and Gulf sturgeon unlikely.

Endangered Species Act consultation with NOAA's National Marine Fisheries Service (NMFS) was conducted for this project. NMFS identified potential effects to sea turtles, smalltooth sawfish, and Gulf sturgeon and concluded that they are not likely to be adversely affected by the proposed project (NMFS consultation letter, April 2, 2012).

Fish
Increases in boating opportunities and recreational fishing are not expected to adversely impact fish populations. The number of new trips generated by the construction and modification of these four boat ramps will not be significant in the context in the total number of trips generated by all access points in Florida.

Birds
The boat ramps would be constructed on already developed sites where it is not likely that nesting shore- and seabirds would be impacted. Contractors are required to be aware of, and comply with applicable law prohibiting harm to migratory birds and endangered species and that the appropriate wildlife permits are obtained if needed.

No Action
No action would result in no changes to biological species.

4.7.4 Cultural Resources

Affected Environment

The area of potential effect used during reviews under Section 106 of the National Historic Preservation Act includes the areas of direct and indirect impact. At the Mahogany Mill new ramp location, some nineteenth century industrial works remnants, a large gear and an old concrete foundation, would be incorporated into the site with educational signage. Ship wrecks and their associated artifacts are historical cultural resources that could potentially be affected in the project area.

Environmental Consequences

A complete review of this project under Section 106 of the National Historic Preservation Act will be completed prior to project implementation. To date, the Florida Division of Historical Resources reviewed the Mahogany Mill, Navy Point and Galvez Landing sites. A cultural resource reconnaissance survey is required for the Perdido River site and has been initiated (February 9, 2012 and December 16, 2011, Florida Department of State, Division of Historical Resources.)

In the event that any cultural resources or human remains are found during construction, all activities involving subsurface disturbance in the immediate vicinity of such discoveries must cease, and the appropriate authorities contacted (December 16, 2011 and February 9, 2012, Florida Department of State, Division of Historical Resources.)

No Action

No impacts to cultural resources are predicted if no action is taken.

4.7.5 Geology, Soils, and Sediments

Affected Environment

There are no anticipated adverse impacts to local geology, soils, and sediments. Sediments at all four proposed locations are primarily sandbottom.

Environmental Consequences

There are no anticipated adverse impacts to local geology, soils, and sediments associated with building on these sites. See Section 4.7.11, Water Resources, for erosion mitigation measures.

No Action

If no action is taken there would be no impacts to geology, soils or sediments of the sites.

4.7.6 Land Use

Affected Environment

The land use is recreational boat launching on bay and river sites. The new boat ramps are proposed in developed areas near industrial, residential, and/or commercial buildings.

Environmental Consequences

Building and establishing boat ramps is consistent with the current land uses for the four building site locations. General land use patterns would not be affected if these projects are implemented.

No Action

If no action is taken, the land use for the building site locations would remain the same.

4.7.7 Noise

Affected Environment

The areas already have boat traffic creating noise with minimal impacts to the wildlife and people in the area. There may be wildlife living near the boat ramp locations which could be impacted by the noise. No residential properties are directly adjacent to any of the new ramp locations.

Environmental Consequences

Machinery and equipment used during construction would generate noise. This noise may disturb wildlife and humans using the area but would be kept to a minimum using best management practices. Once built, the proposed project would not cause long-term noise impacts. There may be minimal noise impacts associated with increased boat traffic on the river and increased vehicle traffic at the ramps. The amount of vehicle traffic at the ramps will not cause long-term noise impacts.

No Action

If no action is taken, there would be no changes in noise.

4.7.8 Socioeconomics and Environmental Justice

Affected Environment

Specific locational information for each site is detailed in Section 4.7.

Environmental Consequences

Local businesses in surrounding areas may benefit from customers utilizing the boat ramps. The Trustees have evaluated environmental justice concerns regarding the project. Based on the overall minimal environmental impact of the project, the proposed project would not create a significant adverse environmental impact on any community or group of people. Therefore there would be no disproportionate share of adverse environmental impacts on any minority, low income, disadvantaged, or Native American tribal population within the area of the proposed project.

No Action

Socioeconomics and environmental justice would not be impacted if the project were not implemented.

4.7.9   Public Access/Recreation

Affected Environment

Boating on the Perdido River and in Pensacola Bay is already a common recreational activity. The Florida Department of Highway Safety and Motor Vehicles records for 2010 indicate more than 16,000 vessels of the size that use boat ramps registered in Escambia County. Escambia County has less than 400 ramp spaces at city/county public ramps for these boats. It is estimated that they only have ramp access for less than 3 percent of registered boats that can be expected to need ramp access. This does not include boats brought into Escambia County by tourists on trailers, which is a high use (pers. comm, Escambia County Marine Resources Division Manager.)

The Marine Advisory Council (MAC) was established by the Escambia County Board of County Commissioners to provide input to Escambia County regarding marine and aquatic resources. Subsequent to the MAC, Escambia County established the Marine Resources Division (MRD) to provide for direct management of marine and aquatic resources. MAC input is one of the sources of information taken into account in public waterway access determinations.

The MRD has been seeking funding and property to site boat ramps since the Division was established in 2000. Public demand for boating and waterways access has exceeded existing resources for many decades, and the MRD received numerous requests to increase parking at existing boat ramps as well as to establish new boat ramps. Perdido Bay has only two small boat ramps. One ramp (Coronada Street) has no parking; the other ramp (Heron Bayou) has parking for only 3-5 vehicles (depending on size/parking pattern). A former boat ramp on private property at Hurst hammock was closed after the property was sold after Hurricane Ivan, eliminating public access to the upper Perdido Bay/Lower Perdido River. This susceptibility for loss of public waterways access is a serious threat because private property owners have eliminated public access to their property for various reasons.

In 2007, an intensive search for additional boat ramp sites in Escambia County was begun. At that time, the Marine Advisory Committee designated Perdido Bay as a high priority due to the facts discussed in the previous paragraph. In August 2011, the MAC unanimously moved to recommend development of the Perdido River public boat ramp a high priority (pers. comm., Escambia County Marine Resources Division Manager).

Environmental Consequences

Access to the Galvez Landing ramp would be slowed for some of the duration of construction. Ramp access at Navy Point would not diminish due to construction. Recreational access would be increased after construction at the new Mahogany Mill and Perdido River locations. The project would improve access to public waterways, benefitting recreational opportunities. There are several reports from the Florida Boating Access Facilities and Economic Assessment that documents the value of boating in your area (http://www.myfwc.com/media/1162807/Registered-BoaterSpending.pdf).

No Action

Recreational access to the Mahogany Mill and Perdido River locations would not be created if the ramp construction at these sites does not take place.

4.7.10 Utilities and Public Services

Affected Environment

Utilities and public services in the project areas would continue to be available throughout the duration of construction and after completion.

Environmental Consequences

The Mahogany Mill ramp is expected to increase vehicular traffic; however, road improvements adjacent to the ramp site are scheduled that will account for the increased traffic due to the enhanced boat ramp. The Perdido River ramp is expected to increase vehicular traffic, but not beyond the current capacity of the surrounding roads.

No Action

There would be no changes to utilities or public services.

4.7.11 Water Resources

Affected Environment

The environment consists of coastal, estuarine, and riverine habitats. More specific locational information for each individual site is detailed in Section 4.7.

Environmental Consequences

With required mitigation in place, impacts to water quality are expected to be minimal. All permit conditions requiring mitigation measures for siltation, erosion, turbidity and release of chemicals will be strictly adhered to. During construction, best management practices and boom placement along with other avoidance and mitigation measures required by state and federal regulatory agencies would be employed to minimize any water quality and sedimentation impacts. The Florida DEP permit conditions require erosion and turbidity mitigation measures. These include:

- Install floating turbidity barriers
- Install erosion control measures along the perimeter of all work areas
- Stabilize all filled areas with sod, mats, barriers or a combination
- If turbidity thresholds are exceeded the project must stop, stabilize the soils, modify the work procedures, and notify the Florida DEP

The Florida DEP permits also constitute a Certification of Compliance with State Water Quality Standards under Section 401 of the Clean Water Act, which means that the project will comply with state water quality standards and other aquatic resource protection requirements.

After construction, increased boat traffic on the river could result in minimal impacts to surface water quality. Boat wakes created by additional boat traffic that could increase shoreline erosion would be controlled through no-wake or speed zones to mitigate shoreline erosion on the river.

The Mahogany Mill ramp construction includes shoreline stabilization by planting native vegetation along the shoreline at this site.

Impacts from chemicals that could potentially be released from sources such as construction equipment and boats are expected to be negligible. Required spill containment measures would be implemented for applicable construction activities. Florida DEP permits require spill containment protection and mitigation measures such as:

- No boat repair or fueling facilities over the water,
- No vessel removed from the water for purposes of maintenance or repair,
- Prohibited activities include hull cleaning and painting, discharges or release of oils or greases, and related metal-based bottom paints associated with hull scraping, cleaning, and painting (Consolidated Wetland Resource Field Permit and Sovereign Submerged Lands Authorization, Florida DEP, July 12, 2010).

No overboard discharges of trash, human or animal waste, including fish carcasses is allowed at the piers (Consolidated Wetland Resource Field Permit and Sovereign Submerged Lands Authorization, Florida DEP, July 12, 2010).

This project would not impact groundwater.

No Action

There would be no adverse impacts to water resources.

4.7.12 Cumulative Impacts

This project would build two new boat ramps in developed areas and enhance two existing boat ramps, providing boaters with enhanced access to public waterways within Pensacola Bay, Perdido Bay and offshore areas. The boat ramp project is not anticipated to create significant adverse environmental or socioeconomic cumulative impacts. Both new construction and existing ramps are located within developed areas; construction and operation would cause only minimal alternation/impacts to habitats. The boat ramp project is not expected to have significant adverse cumulative impact on manatees because the ramps are on the fringe of the manatee's habitat range. Where applicable, manatee education and restrictions necessary to protect manatees during boat ramp construction will be implemented. Manatee education signs and manatee educational programs for operation of boat ramps would be installed, where applicable.

Increased boating opportunities and recreational fishing are not expected to adversely impact fish populations. The number of new trips generated by the construction and modification of these four boat ramps will not be significant in the context of the total number of trips generated by all access points in Florida.

The boat ramp project would not have significant adverse cumulative impacts to wildlife habitat, fisheries, threatened or endangered fish species, cultural resources or other resource areas. Although the Florida boat ramp early restoration project has potential short-term negative effects, on balance, the proposed project has positive effects that are consistent with long-term planning goals, and contribute beneficially to access and aquatic recreation in the Florida panhandle. Additionally, all immediate effects are relatively local and geographically disparate.

The Florida boat ramp project will adhere to all applicable permit conditions, federal, state and local requirements for the protection of marine mammals and other environmental resources during the construction and operation of the project.

No Action

Under the No Action alternative, there would be no adverse cumulative impacts to natural resources.

4.7.13 Summary

Overall, this proposed project would enhance aquatic recreation activities on the Perdido River and in Pensacola Bay. The beneficial public access effects are expected to far outweigh any short-term, adverse impacts from construction or operational impacts. Impacts to coastal, marine, estuarine, and riverine biological resources due to increased human activity are expected to be minimal. Implementation of the proposed project should not result in substantial impacts to water quality. In summary, the Trustees believe that the proposed project in this restoration plan will not cause substantial adverse impacts to natural resources or the services they provide. Furthermore, the Trustees do not believe the proposed projects will affect the quality of the human environment in ways deemed substantial.

4.8 Florida (Pensacola Beach) Dune Restoration

Purpose and Need

Florida dunes were exposed to oil, dispersants and/or affected by response activities undertaken to prevent, minimize, or remediate oiling from the Spill. Under OPA, the Trustees act on behalf of the public to restore, rehabilitate, replace, or acquire the equivalent of natural resources injured and associated service losses as a result of the Spill. Under the Framework Agreement, the Trustees have the opportunity to negotiate with BP to fund early restoration projects. The purpose of a Florida dune restoration project implemented under OPA and the Framework Agreement is to meet the need to restore Florida dune resources.

General Project Information

Dune vegetation in the Pensacola Beach area of Escambia County, which is a coastal area in the western panhandle of Florida, has been injured by exposure to DWH oil and/or response activities undertaken to prevent, minimize and remediate oiling. The project is needed to help restore an area of the beach where oiling and the heavy use of excavators, tractors, trailers, ATVs, and other equipment on beaches resulted in the trampling and removal of sand, vegetation, wrack, and shell which has inhibited plant growth and prevented the seaward expansion of dunes since June 2010. This project would provide restoration of the dune profile and replace vegetation injured or destroyed by response activities, as well as decreasing erosion in the area.

The primary dunes are the first natural line of defense for coastal Florida to prevent the loss of wildlife habitat and private property due to hurricanes, sea level rise, oil spills, and other threats. The State proposes to restore beach and dune habitats in Florida that were affected by the Spill by planting native primary dune vegetation.

The proposed dune restoration project would help prevent beach erosion by restoring a "living shoreline," a coastline protected by plants and natural resources rather than hard structures. As a result, this project would assist with the restoration of wildlife, habitats, and communities along the northwest Florida Gulf Coast. Project details include:

- No new access roads or staging areas would be built as part of this project.
- Vehicles would use existing roads and parking areas.
- All participants involved in the project would follow rules established to minimize foot traffic and human presence across ecologically sensitive areas.
- No threatened or endangered species, or eligible cultural sites or historic properties would be negatively affected as a result of implementing this project. There are no endangered or threatened beach mouse species in the affected area, and due to the narrowness of the beach, there is no nesting of endangered or threatened shorebirds (piping plovers, least terns) along that segment of beach. Some piping plover winter habitat falls in the project area.
- The planting portion of the project would occur during the growing season (approximately March-August). Care would be taken to ensure plants would be installed in areas where nesting sea turtles (primarily loggerhead; Kemp's Ridley, leatherback, green are rare to occasional nesters) and shorebirds would not be impacted. All plants would be grown from seeds or cuttings from the Alabama coast or North Florida to ensure appropriate genetic stocks are used in the project.
- Installation of sand fencing is not part of this project.

NEPA compliance

NEPA requires Federal agencies to analyze their proposed actions to determine if they could have significant environmental effects. Over time, through study and experience, agencies may identify activities - such as routine facility maintenance - that do not need to undergo detailed environmental analysis in an environmental assessment (EA) or an environmental impact statement (EIS) because the activities do not individually or cumulatively have a significant effect on the human environment. Agencies can define categories of such activities, called categorical exclusions (CXs), in their NEPA implementing procedures, as a way to reduce unnecessary paperwork and delay.

If an agency determines that a proposed activity fits within the description of a categorical exclusion and that there are no extraordinary circumstances that might cause significant environmental effects, no additional NEPA review is required and the agency can proceed with the activity without preparing an EA or EIS. Categorical exclusions are an essential tool in facilitating NEPA implementation and concentrating environmental reviews on instances of potential impacts. (CEQ issued NEPA Guidance on Categorical Exclusions on November 23,

2010.) A CX is a form of NEPA compliance, without the analysis that occurs in an EA or an EIS. It is not an exemption from the NEPA. The Departmental Manual (516 DM 2.3A (3) and 516 DM 2, Appendix 2) requires that before a CX is used the list of "extraordinary circumstances" be reviewed for applicability. When no "extraordinary circumstances" exist, neither an EA nor an EIS is required (40 CFR 1508.4).

After undergoing NEPA review, the Trustees determined this project would meet two resource management categorical exclusions as described in in the U.S. Fish and Wildlife Service's Service NEPA Procedures in Departmental Manual (516 DM6 Appendix 1, Section 1.4, nos. 3 and 11) and BLM Department Manual 516 DM 11.9. These categorical exclusions are:

(3) The construction of new, or the addition of, small structures or improvements, including structures and improvements for restoration of wetland, riparian, instream, or native habitats, which result in no or only minor changes in the use of the affected local area. The following are examples of activities that may be included.
i. The installation of fences.
ii. The construction of small water control structures.
iii. The planting of seeds or seedlings and other minor revegetation actions.
iv. The construction of small berms or dikes.
v. The development of limited access for routine maintenance and management purposes.

(11)Natural resource damage assessment restoration plans, prepared under sections 107, 111, and 122(j) of the Comprehensive Environmental Response Compensation and Liability Act (CERCLA); section 311(f)(4) of the Clean Water Act; and the Oil Pollution Act; when only minor or negligible change in the use of the affected areas is planned.

Because the dune restoration project involves planting and other minor revegetation activities, it would result in only minor or negligible change in the use of the project area, U.S. Fish and Wildlife Service determined that it would apply the categorical exclusion iii described above to this project.

A NEPA Compliance Checklist and Environmental Action Statement (EAS) were prepared to document the use of the categorical exclusions. An EAS is "a Service-required document prepared to improve the Service's administrative record for categorically excluded actions that may be controversial, emergency actions under CEQ's NEPA regulations (40 CFR 1506.1 1), decisions based on EAs to prepare an EIS, and any decision where improved documentation of the administrative record is desirable, and to facilitate internal program review and final approval when a FONSI is to be signed at the FWS-WO and FWS-RO level (550 FW ')".

Measures that would ensure there would be no negative effect on sea turtles include adhering to the following Florida DEP's Coastal Construction Control Line Special Permit Conditions for Dune Planting within Sea Turtle Nesting Season:

1. It is the responsibility of the permittee to ensure that the project area and access sites are surveyed for sea turtle nesting activity. All nest surveys, nest relocations, screening or caging activities etc. shall be conducted only by persons with prior experience and training in these

activities and is duly authorized to conduct such activities through a valid permit issued by FWC the pursuant to Florida Administrative Code 68E-1. For information regarding whether the project beach is surveyed by qualified personnel, contact FWC at 561/575-5407.

2. Sea turtle nest surveys shall be initiated at the beginning of the nesting season or 65 days prior to installation of plants (whichever is later). Surveys shall continue until completion of the project or through September 15 (whichever is earliest). Surveys shall be conducted throughout the project area and all beach access sites.

3. Any nests deposited in an area not requiring relocation for conservation purposes (as determined by the FWC authorized marine turtle permit holder) shall be left in situ. The marine turtle permit holder shall install an on beach marker at any nest site and a secondary marker located at a point as far landward as possible to ensure that future location of the nest will be possible should the on-beach marker be lost. A series of stakes and survey ribbon or string shall be installed to establish an area of 3 feet radius surrounding the nest. No planting or other activity shall occur within this area nor shall any activity occur which might cause indirect impacts within this area. Nest sites shall be inspected daily to ensure nest markers have not been removed.

4. The use of heavy equipment (trucks) is not authorized seaward of the dune crest or armoring structure. A lightweight (ATV style) vehicle, with tire pressures of 10 p.s.i. or less can operate on the beach.

5. Any vegetation planting and removal or placement of irrigation materials shall be conducted with hand labor and tools.

6. Irrigation (if proposed) shall be entrenched 1 to 3 inches below grade so as not to pose a barrier to sea turtle hatchlings and to allow for easy removal. Irrigation piping shall avoid all marked nests by a minimum of ten (10) feet. The irrigation system shall be designed and maintained so that watering of the unplanted sandy beach does not occur. In the event a sea turtle nest is deposited within the newly established dune planting area, the permittee shall modify the irrigation system so that watering within 10 feet of the nest does not occur. Daily inspection of the irrigation system shall be conducted by the permittee to ensure compliance with this condition.

7. All activity shall be confined to daylight hours and shall not occur prior to the completion of all necessary sea turtle surveys and conservation activities within the project area. Nighttime storage of equipment or materials shall be off the beach (landward of the dune crest, existing seawalls or bulkheads).

8. In the event a nest is disturbed or uncovered during planting activity, the permittee shall cease all work and immediately contact the person(s) responsible for sea turtle conservation measures within the project area. If a nest(s) cannot be safely avoided during construction, all activity within the affected project area shall be delayed until complete hatching and emergence of the nest.

Endangered Species Act consultation has been completed for this project (USFWS, Intra-service Section 7 Biological Evaluation, February 7, 2012.). All recommended mitigation measures to protect these species will be implemented.

A complete review of this project under Section 106 of the National Historic Preservation Act will be completed prior to project implementation. To date, the Florida Division of Historical Resources reviewed the Florida (Pensacola Beach) Dune Restoration Project and determined, due to the of the nature of the project, it is unlikely that historic properties will be affected (Florida Department of State, Division of Historical Resources, February 23, 2012).

Since project scopes, environmental conditions and regulatory requirements can change over time, the use of these CXs would be reviewed for their continued applicability to the project before implementation.

Summary

No threatened or endangered species, or eligible cultural sites or historic properties would be affected as a result of implementing this project. ESA Section 7 consultation has been completed and the Trustees do not anticipate an adverse impact to sea turtles or nesting shorebirds. Endangered beach mice do not occur in the project area. A complete review of this project under Section 106 of the National Historic Preservation Act will be completed prior to project implementation.

Overall, this project would enhance the Pensacola Beach dune ecosystem. The Trustees determined that the proposed activity qualifies for two categorical exclusion(s) and that there are no extraordinary circumstances that might cause significant environmental effects. Further, the Trustees believe the project would have no potential adverse impact on the quality of the human environment, either individually or cumulatively. Accordingly, no additional NEPA analysis for this project is required at this time.

CHAPTER 5 PUBLIC COMMENT ON DRAFT PHASE I EARLY RESTORATION PLAN AND ENVIRONMENTAL ASSESSMENT AND RESPONSES

The public comment period for the DERP/EA opened December 14, 2011 and closed February 14, 2012. During this time, the Trustees hosted 12 public meetings in Texas, Louisiana, Mississippi, Alabama, Florida and Washington D.C., at which the Trustees accepted written comments, as well as verbal comments recorded by court reporters. The Trustees also hosted web-based comment submission sites, as well as provided a P.O. Box and email address during the comment period with which to receive comments. In addition to comments provided at public meetings, the Trustees received web-based submissions, emailed submissions, and mailed-in submissions.

Following the comment period, the Trustees reviewed all submissions. Similar or related comments were then grouped and summarized for purposes of response. All comments submitted during the period for public comment were reviewed and considered by the Trustees prior to finalizing the Phase I ERP/EA. All comments submitted are represented in the summary comment descriptions listed in this chapter.

Comments received were both general in nature as well as directed toward specific aspects of one or more of the eight projects. In addition, two larger manuscripts were submitted as part of the public comment period. These manuscripts are noted and addressed individually. All public comments will be included in the AR.

5.1 General Comments

Comments that were not specific to particular projects, but generally applicable to the public comment process, project selection, residual contamination, project implementation, monitoring, new project ideas and other 'general' topics are addressed in Section 5.1. Comments specific to particular projects are addressed in Section 5.2.

Comment: A Native American community asked the Trustees to work collaboratively with them to restore resources important to them and potentially injured by the DWH spill.
Response: The Trustees have and will continue to engage any and all interested stakeholders, such as Native American communities, through public outreach and coordination. For example, in addition to a variety of public meeting settings, the Louisiana Trustees frequently meet with stakeholders, including Native American communities, both individually and collectively, to discuss NRDA, the restoration planning process and potential restoration project ideas specifically related to the Spill.

Comment: Comments suggested other potential restoration projects.
Response: The Trustees will continue to evaluate new and existing project ideas as potential DWH NRDA restoration projects. Project ideas can continue to be submitted and reviewed at http://www.gulfspillrestoration.noaa.gov/restoration/.

Comment: Some commented that more Gulf of Mexico information should be presented in Chapter 2.

Response: The Trustees understand the interest in more detailed information about the Gulf but believe the information presented in Chapter 2 is sufficient given the purpose of this information within this plan. The intent of Chapter 2 is to describe the general environment of the Gulf that provides the setting for the restoration projects and the resources or services expected to benefit from those projects. Additional information on the environmental setting for each proposed early restoration project is also included in Chapter 4.

Comment: Some commenters expressed general support for the early restoration process and projects proposed in Alternative B.

Response: The Trustees acknowledge this support.

Comment: Some commenters made a general request for additional information about the projects proposed in Alternative B.

Response: The DERP/EA included a considerable amount of information about the projects as well as the context and basis for their selection under OPA and the Framework Agreement. The Trustees believe the information is sufficient to inform the public about these early restoration proposals and to provide meaningful comment on the proposed projects.

Comment: The DERP/EA was not adequate in compensating the public for injuries incurred from the spill and response activities.

Response: The projects proposed in the DERP/EA represent only the first projects in the earliest phase of the DWH NRDA restoration process. Injury assessment and restoration planning are ongoing. The Trustees continue to evaluate additional projects for funding as part of the early restoration process but also to work toward developing longer term restoration plans with the goal of fully compensating the public for all resource injuries and losses that resulted from the Spill.

Comment: The Trustees should examine the socio-economic impacts of individual restoration projects to analyze both the potential benefits to the most impacted communities and any potential negative costs to the Gulf's low income, indigenous, and disadvantaged populations.

Response: The Trustees considered whether the proposed projects will result in adverse human health or environmental effects in low income, indigenous and disadvantaged communities on a project-by-project basis. The Trustees found no indication that any such population would be affected by any anticipated adverse environmental impacts associated with any proposed projects.

Comment: Concern was expressed that none of the Environmental Assessments for the Alternative B proposals included an assessment of impacts to marine mammals.

Response: The DERP/EA includes evaluations regarding potential project impacts to marine mammals, but only for those projects that might affect marine mammals. Additionally, following publication of the DERP/EA, potential project impacts on marine mammals were also evaluated and addressed through ESA Section 7 consultations with appropriate federal agencies. EA sections for projects with any potential to impact marine mammals have been updated in this ERP/EA to reflect information obtained and the outcome or status of those efforts.

Comment: Commenters suggested that a full EIS should accompany the DERP.
Response: The Trustees disagree that preparation of an EIS is required to provide for sufficient environmental analyses of all eight projects, as explained in Chapter 4 and consistent with CEQ guidance found in 40 C.F.R. § 1508.25.

Comment: Commenters requested additional evaluation of cumulative impacts.
Response: The Trustees updated the cumulative impact section of the ERP/EA, and believe that the information presented meets NEPA requirements.

Comment: A final Phase I ERP/EA should incorporate climate change adaptation measures to ensure that Gulf restoration is focused on creating a more resilient future.
Response: Environmental changes, such as anticipated sea level rise, have been or will be factored into project designs, where appropriate.

Public Comment Process

Public Comment Process comments addressed several types of process-related issues, including, but not limited to, the public meeting format, requests for additional meetings, the potential need for an additional comment period and Trustee consideration and integration of comments.

Comment: Comments addressing the public meeting process included thanks and support for the public meetings and comment process as well as requests for additional meeting locations, times and logistic changes (e.g., regarding language translation).
Response: The Trustees recognize that public input is a critical part of the NRDA early restoration planning process. Official announcements of the public comment period were published in the Federal and Louisiana Registers. The Trustees also hosted a series of 12 public meetings during the public comment period in Texas, Louisiana, Mississippi, Alabama, Florida and Washington D.C. to directly facilitate opportunities for public input. The Trustees believe the number and location of meetings was appropriate. A variety of mechanisms were employed to make the public aware of meeting locations and times in advance, including but not limited to notice and information on the Trustee websites and in newspapers across the Gulf coast. Translation services were provided at public meetings where a need for such services was anticipated or requested.

Comment: Change the format of public meetings to allow the public to provide comments on proposals prior to holding a meeting so that feedback can be discussed at meetings.
Response: The combination of open house and listening session format of the meetings was used to provide an avenue for the public to review and comment on the proposed draft ERP/EA if they had not already done so. Immediately prior to and following each listening session in which the public was encouraged to comment, subject matter experts and Trustee representatives were available to discuss issues related to the NRDA and early restoration proposals. These sessions provided an avenue for one-on-one discussion and feedback.

Comment: A second round comment period should be incorporated into the Phase I DERP/EA process or future coastal restoration public comment procedures to ensure that final plans incorporate public input.

Response: The Trustees concluded that a second notice and an additional comment period was not needed. This decision reflects consideration of the scope of this plan (eight projects), the length of the first public comment period (60 days), the variety of means provided for public input, the number and comprehensive nature of the public comments submitted on the DERP/EA, and the resulting nature of changes made to the DERP/EA, among other factors. Chapter 5 summarizes the Trustees' review and incorporation of public comment into the ERP/EA.

Comment: Commenters questioned how public comments are integrated and considered.
Response: All comments submitted during the period for public comment were reviewed and considered by the Trustees prior to finalizing the Phase I ERP/EA. For efficiency and to aid in timely review, similar or related comments were grouped and their merits and implications considered. The ERP/EA outlines the Trustees' decision regarding moving forward with the early restoration process as to the eight proposed Phase I projects, after taking into account the public comment on the DERP/EA. This Chapter summarizes the comments for purposes of response and describes the results of their consideration by the Trustees.

Planning and Project Development

Planning and project development comments included, but are not limited to, comments on how projects fit into Gulf planning initiatives, project selection criteria, types of projects, questions on why Texas did not have a project in Phase I and why Florida had the only human use project.

Comment: General requests were made for information on how specific projects fit into regional restoration plans and/or an overall Gulf of Mexico restoration strategy.
Response: Restoration projects and strategies developed under regional plans, such as Louisiana's Coastal Master Plan and Annual Plan updates and the Mississippi Coastal Improvements Program, as well as those developed through Gulf-wide efforts, such as the Gulf Coast Ecosystem Restoration Task Force (GCERTF, 2011), Mabus (2010), Brown et al. (2011), NRCS (2011) and Peterson et al. (2011), have been and will continue to be considered in the early restoration process. Additional information on the relationship of specific projects to particular plans or strategies is included in Section 5.2 of this chapter.

Comment: Comments suggested that comprehensive restoration of the Gulf of Mexico, including addressing issues such as long-term sustainability, contaminants of concern and the annual development of a zone of hypoxia (also known as "the dead zone"), should be considered.
Response: The Trustees are mindful of the full array of regional environmental issues in the Gulf region. In undertaking planning for restoration actions, the Trustees have and will continue to consider actions which address regional or Gulf-wide issues that are consistent with the purposes and goals of restoration under OPA (i.e., compensate the public for losses of natural resources and services resulting from the spill). The Trustees also have and will consider potential impacts of those issues on restoration plans being developed.

Comment: A document titled *A Once and Future Gulf of Mexico Ecosystem: Restoration Recommendations of an Expert Working Group* was submitted for the record. This document provides multiple suggestions on a large scale, long-term GOM restoration strategy, referencing other similar documents.

Response: The Trustees have and will consider recommendations provided in this document, along with other Gulf-related restoration documents, in the restoration planning process.

Comment: A document titled *Sunshine on the Gulf: the case for transparency in restoration project selection* was submitted for the record. This document raises concerns regarding the transparency of the Trustees' early restoration project evaluation and selection process, their negotiations with BP, and included a proposed model for evaluating potential restoration projects.[24]

Response: The Trustees have been forthcoming regarding their approach to evaluating and selecting projects in the early restoration process and believe the DERP/EA provided sufficient information in that regard. Their goals in the early restoration process, the project selection criteria that they applied, and their reasons for proposing the projects included in Alternative B are clearly articulated in the DERP/EA. The process the Trustees followed is consistent with applicable laws, regulations and the publicly available Framework Agreement. The Phase I projects were identified through a reasonable balancing of early restoration project objectives, opportunities and timelines in the process of applying project selection criteria.

With respect to the negotiation process, as discussed in the DERP/EA, under the Framework Agreement each early restoration project is subject to negotiation with BP and agreement on project costs, BP funding and NRD Offsets. Initial negotiations were conducted with BP as a means of determining whether agreements-in-principle on the Trustees' proposed projects were achievable prior to preparing the DERP/EA. Such initial agreements, however, are subject to the outcome of the public review of the proposed projects as presented in the DERP/EA. For projects selected for the ERP/EA, the negotiated agreements on costs, funding and NRD Offsets will be included in the Administrative Record in accordance with the terms of the Framework Agreement.

Comment: The Trustees were asked to provide project evaluation criteria to help guide future proposals.

Response: The project selection criteria are identified in the DERP/EA. The Trustees will continue to use these criteria to evaluate early restoration proposals.

Comment: Comments were submitted on project selection criteria and on scoring of restoration proposals. Requests were made to see more information on criteria evaluations for projects and supporting evaluation documents.

Response: The Trustees utilized project selection criteria as identified in the DERP/EA. These criteria are consistent with applicable laws, regulations and the Framework Agreement. As noted above, the Phase I projects were identified through a reasonable balancing of early restoration project objectives, opportunities and timelines in the process of applying project selection criteria. While individual Trustees considered projects, proposed projects were selected and included in the DERP/EA based on consensus of all the Trustees using the same selection criteria. The DERP/EA provides more detailed information regarding early restoration project selection.

---

[24] The report includes additional comments regarding public participation and project evaluation and selection that are like other comments summarized and addressed herein.

Comment: Concern was raised about selecting restoration projects prior to completing an injury assessment.

Response: The Trustees continue to assess the injuries and losses of natural resources and services caused by the DWH spill. The Spill event, however, was extraordinary and some impacts to species, habitats, and resource uses were manifest early on across the Gulf. The Trustees feel strongly that it is in the best interest of the public and the environment to take early steps to accelerate the restoration process in this instance and to begin implementing projects that help restore and/or compensate for those losses even before the completion of the full damage assessment.

Comment: It was suggested that the Trustees only approve projects that legitimately and effectively address natural resource damages, that follow all environmental laws, and that are science-based in their assessment and implementation.

Response: As described in the DERP/EA, the Trustees adhere to OPA, NEPA, other applicable laws and regulations, and use the best available science in planning for early restoration. All necessary environmental compliance activities will occur and any requirements will be met before a project is implemented. To be part of early restoration, a restoration project must be capable of restoring, replacing, rehabilitating or providing resources or services equivalent to those lost, injured or destroyed as a result of the Spill. The nexus to injury is a threshold project selection criterion under OPA and ensures that restoration projects will appropriately and effectively contribute to addressing natural resource damages.

Comment: Some comments did not support projects such as piers, renourishment of beaches with dredge or fill material, or coastal armoring that degrade the natural beach environment.

Response: The DERP/EA included two oyster projects, two marsh projects, a nearshore artificial reef project, two dune planting projects, and a boat ramp project. None of these projects will result in significant degradation of a natural beach environment. Any adverse impacts during implementation will be negligible and of short duration. Any potential projects that may be considered for early restoration will be evaluated using the same project selection criteria as identified in the DERP/EA and consistent with applicable laws, regulations and the Framework Agreement.

Comment: DWH NRDA restoration should be large-scale, sustainable, and adaptive to subsidence and climate impacts, promote projects that have the potential to enhance each other, and follow an overall strategy for Gulf of Mexico recovery.

Response: The goal of the NRDA regulations within OPA is to make the environment and public whole for injuries to natural resources and associated services resulting from an oil spill incident. The Phase I ERP/EA represents the first step on the road to a full recovery for the region following the Spill. As discussed in the DERP/EA, the Trustees used a number of selection criteria to evaluate project proposals. The Trustees believe projects selected in the ERP/EA are consistent regional restoration planning efforts in the Gulf, which take into account factors such as potential subsidence and climate change impacts, and promote Gulf-wide enhancement.

Comment: A number of commenters asked for additional transparency in the Trustee DWH NRDA restoration process.

Response: The Trustees understand the importance and value of transparency in the NRDA restoration process and made substantial efforts to ensure the public is aware of the goals of restoration, the criteria to be applied in choosing restoration projects under OPA, the on-going opportunities for the public to submit projects for consideration, and the terms and processes outlined in the Framework Agreement that must also be satisfied to access BP funding. The Trustees have held numerous public meetings and developed and actively manage several web-based information portals used to keep the public apprised about restoration planning for the DWH spill. The Trustees will continue to look for ways to improve their efforts in this regard, provided this can be accomplished consistent with timing, resource, cost and legal constraints.

Comment: Requests were made for additional public participation and/or for the development of a citizen's advisory group to facilitate public interaction with the Trustees.

Response: Public input is a critical part of the NRDA early restoration planning process. As it relates to early restoration planning, the Trustees have invited or solicited public input through a variety of mechanisms, including, but not limited to: 1) development and management of several web-based information portals, 2) multiple preliminary public meetings to educate the public on the process and solicit input; 3) active solicitation of restoration project proposals from the public; 4) hosting of 12 public meetings on the DERP in Texas, Louisiana, Mississippi, Alabama, Florida and Washington D.C.; and 5) two months for public review and comment on the Phase I projects proposed in the DERP. Collectively, the opportunities afforded the public to participate in early restoration planning have been substantial and more extensive than those afforded in NRDA processes for other oil spills. The enhanced efforts for the *Deepwater Horizon* Spill are viewed by the Trustees as commensurate with the nature of the Spill. An administrative record has been established, providing an additional mechanism for the public to review case-related information and documents.

The Trustees understand and value the public interest in early restoration, and strive to maintain a high degree of transparency while protecting the integrity of the case and fulfilling the critical mission to protect, preserve, and restore the Gulf's natural resources. The Trustees have and will continue to provide ample opportunities for all members of the public to provide input into the early and longer term restoration planning processes, including ongoing activities regarding public stakeholder groups. At this point, the Trustees have no plans to form a special citizen's advisory group for the DWH NRDA early restoration process.

Comment: The Trustees preselected projects prior to public comment period.

Response: The Trustees actively solicited public input through a variety of mechanisms for potential restoration projects prior to development of the Phase I DERP/EA. Consistent with the Framework Agreement, each Trustee prioritized project proposals based on established selection criteria prior to bringing them forth for Trustee group consideration and initial negotiation with BP on cost and Offsets. The DERP/EA represented the Trustees' proposal of an initial list of projects. The Trustees considered all public comments on the DERP/EA prior to finalizing the selection of the Phase I early restoration projects in the ERP/EA.

Comment: The Phase I early restoration plan should include a project in Texas.

Response: The Texas Trustees have been carefully looking at the potential injuries and losses to Texas' trust resource interests as part of the early restoration planning process. The Texas

Trustees are working to identify the projects most suitable to meet early restoration goals for injuries to its resource interests. Texas will propose projects for inclusion in future early restoration plans.

Comment: A question was submitted as to why Florida was the only state proposing a human use project.

Response: The Trustees sought a diverse set of projects in Phase I to provide benefits to an array of injured resources and lost resource services. While the Florida boat ramp construction and enhancement proposal was the first human use project proposed by the Trustees for inclusion in an early restoration plan, additional human use project proposals are anticipated in subsequent phases of the early restoration process.

Comment: Some commenters opposed selection of human use projects rather than selection of ecological restoration projects.

Response: The goal of the NRDA regulations within the Oil Pollution Act is to make the environment and public whole for injuries to natural resources and associated services resulting from an oil spill incident. The *Deepwater Horizon* Spill caused natural resource injury and the loss of ecological and human uses services. The Trustees will seek compensation for and have a responsibility to consider both types of losses in planning restoration that addresses the impacts caused by the DWH spill.

Comment: Comments requested that the Trustees consider long-term species and ecosystem recovery when selecting projects.

Response: As discussed in Chapter 1, the Trustees used project selection criteria as outlined in OPA and the Framework Agreement together with a number of practical considerations appropriate to the early restoration phase of planning to evaluate and identify the proposed projects. The Trustees agree that the long-term recovery of species and Gulf ecosystems is important to consider in restoration planning processes. Gulf-wide and regional restoration plans consider long-term ecosystem and species recovery and have been and will continue to be considered in the early restoration process.

Comment: The Trustees should design projects that involve young people.

Response: The Trustees recognize the value of incorporating public involvement into DWH NRDA restoration programs and activities, including the involvement of our youth. Implementing Trustees may also seek to include youth in implementing restoration actions where possible, consistent with applicable laws, regulations and policies governing contracting and procurement, laws and policies governing the use of volunteers and safety considerations.

Comment: BP should not be allowed to dictate the selection of restoration projects.

Response: BP is not dictating the selection of restoration projects. The Trustees are fully responsible for the NRDA for the DWH spill, including all decisions on restoration actions that are appropriate to undertake in compensating for all Spill-related injuries and losses of natural resources and uses in the Gulf. The Framework Agreement makes funding available for Trustee-selected projects in return for agreement on Offsets against the Trustees' assessment of natural resource injuries and losses of the public.

Comment: The Trustees should develop and propose Phase II restoration projects swiftly following adoption of Phase I projects.

Response: The Trustees' consideration and planning for additional early restoration projects continues. Additional early restoration projects will be proposed in a Phase II plan as expeditiously as possible.

Comment: Gulf states impacted by the DWHOS should collaborate on regional oyster repopulation efforts that are long-term and scientifically based.

Response: The Phase I ERP/EA includes oyster cultch projects in Mississippi and Louisiana. These two states coordinated extensively in the development of these projects. These projects and associated project designs are consistent with project selection criteria, scientifically based and consistent with existing management practices in proposed project areas. The Trustees have and will continue to coordinate on oyster and other projects to address injuries caused by the Spill.

Project Area Contaminants of Concern

Comment: Residual DWH oil, response actions and other activities and/or sources of contamination in project areas may negatively impact proposed projects. Coordinate restoration with response (clean-up) activities.

Response: Prior to implementing any project the Trustees will coordinate with the Federal On Scene Coordinator to ensure that the project does not obstruct, duplicate, or conflict with any ongoing response actions and that any response actions will not obstruct, duplicate or conflict with the project. Responses specific to projects are provided in later sections of this document. If such issues arise prior to and/or during project implementation, the Trustees may be able utilize contingency funds to modify project design, timing and/or otherwise adaptively manage problems.

Comment: One comment expressed concern regarding continued dispersant use.

Response: The Trustees are evaluating potential injuries to natural resources caused by dispersants used during the Spill. The public is encouraged to contact the Gulf Coast Incident Management Team for information regarding any continued use of dispersants.

Comment: The DERP/EA asserted that dispersants were found in the near shore environment and marshes, and suggested that dispersant exposure may have caused injury to marshes, oysters, and other wildlife in the near shore environment. The purpose of this comment is to ensure that the administrative record shows that dispersants used in response to the DWH spill were not found at levels that equal or exceed any established toxicity threshold near shore or on shore, none of the validated water samples exceeded the toxicity threshold for dispersant marker chemicals, and that the Operational Science Advisory Team found that any dispersants in the water and sediment samples which they reviewed were below government aquatic toxicity benchmarks.

Response: The assessment is ongoing. The Trustees are still assessing the extent to which exposure of natural resources to dispersants may have caused or contributed to any injuries or losses of natural resources or services. While information such as that referred to in the comment may be included in the NRDA Administrative Record, the decision to do so is within the

Trustees' discretion. The NRDA Administrative Record does and will continue to include data that the Trustees rely on in assessing injuries to natural resources, including any injuries related to dispersant use.

Comment: Some commenters expressed general views on injury assessment issues (e.g., type, cause and/or extent of impact), without reference to specific language in the DERP/EA.
Response: The Trustees continue to assess potential injuries and losses to natural resources and services caused by the DWH spill. The Trustees believe that the Phase I ERP/EA includes sufficient information on this topic, consistent with OPA requirements and the Framework Agreement.

Comment: More DWH oil spill cleanup is needed.
Response: Decisions regarding ongoing Spill cleanup are outside the scope of NRDA process. The public is encouraged to contact the Gulf Coast Incident Management Team with any concerns regarding the need for additional or continued response actions for the Spill.

Comment: Continue injury assessment and/or monitoring.
Response: The Trustees are continuing to assess the potential injuries and losses to natural resources and services caused by the DWH spill.

Project Implementation

Comment: How will a project implementation oversight process be conducted?
Response: The Trustees will be responsible for overseeing implementation of all restoration projects. Progress on project implementation will be available to the public.

Comment: Comments were received suggesting financial tracking, auditing and/or reporting of project expenditures.
Response: Financial oversight, auditing and reporting will follow applicable laws, regulations, policies and project-specific legal agreements between the Trustees and BP.

Comment: Requests were made for the Trustees to hire local work forces and to hire people negatively impacted by the spill to implement restoration projects.
Response: The Trustees support this goal in principle, but recognize that implementing Trustees are subject to and must abide by laws, regulations and policies governing their contracting and procurement processes and practices. Such laws, regulations and policies will vary, depending on the Trustee agency implementing a project. Implementing Trustees will be encouraged to give preference to local hiring to the extent permitted by law.

Comment: Implement Phase I projects promptly and efficiently.
Response: Projects will be implemented as quickly and efficiently as possible following publication of the Phase I ERP/EA and finalization of project-specific legal agreements with BP.

Comment: Will there be disincentives for BP delaying funding for projects?
Response: Pursuant to the Framework Agreement, BP has set aside $1 billion in a trust for use exclusively to fund early restoration projects. The fact that BP is unable to use the funds for other

purposes minimizes any incentive for delay. After project-specific agreements have been signed for an early restoration project, BP will be subject to a legally binding schedule for payments. The Trustees will continue to push forward in the early restoration process in an expedient and efficient manner.

Comment: The Trustees should report the number of jobs created by restoration projects as part of transparency and accountability measures.

Response: Although not required by law nor a metric for measuring the success of a project in restoring, replacing, rehabilitating or acquiring the equivalent of resource injuries and losses, the Trustees are open to reporting this type of information if it is available and can be determined or estimated for each project by the implementing Trustees.

Comment: Some commenters provided commercial information relating to products or services potentially relevant to project implementation.

Response: The Phase I ERP/EA is not a solicitation for bids, contractor qualifications or similar information related to the procurement of goods and services needed to implement projects. The Trustees will abide by procurement procedures, consistent with relevant regulations and policy, to address such needs as they arise.

Monitoring

Comment: Information for the Gulf of Mexico is lacking, hampering adequate monitoring efforts.

Response: Monitoring for early restoration projects is focused on the evaluation of project success. Pre- and post-project implementation data will be available and sufficient to meet that objective. The Trustees anticipate developing broader monitoring efforts in later stages of the damage assessment and restoration planning process.

Comment: The NRDA regulations under OPA require that monitoring procedures and metrics for evaluating project performance and triggering response measures are developed and described for each proposed restoration project before environmental review and public comment. The comment referenced 15 C.F.R. Section § 990.55(b)(2).

Response: OPA NRDA regulations set forth several factors that the monitoring component of a Draft Restoration Plan should address to effectively gauge a project's success. Each of the proposed projects in the DERP/EA included a discussion of the performance criteria, monitoring and maintenance plan appropriate for that project. Additional project-specific information has been included in the Phase I ERP/EA in response to public comment. While the details varied somewhat by project, the level of information included is consistent with legal requirements. Additional monitoring information may be developed for some projects. Results of such activities will be made available to the public. All future monitoring plans will be made available to the public.

Comment: General requests were made for more information on project success monitoring and potential adaptive management strategies as they relate to project goals, and for the results of these activities to be shared with the public. Comments recommended that each approved project have sufficient funds budgeted, success benchmarks, and a specified time frame for evaluating

success that supports adaptive management to ensure anticipated project benefits are achieved, adequate compensation for the public and assist in future project implementation efficiency.
Response: As noted above, while the details vary by project, the level of information included in the ERP/EA is consistent with legal requirements. Additional monitoring information may be developed for some projects. These issues are addressed further on a project-specific basis in Section 5.2. Results of such activities will be made available to the public.

Comment: It was suggested that early restoration funds be set aside for a long-term Gulf monitoring program, addressing resources and locations beyond the project-specific monitoring efforts identified in the ERP/EA.
Response: The intent of the early restoration process is to implement projects which accelerate the restoration of resources injured by the DWH spill. Long-term Gulf monitoring, while an important issue, does not meet this objective and is outside the scope of what the Trustees anticipate accomplishing as early restoration under the terms of the Framework Agreement with BP.

Other

Comment: Comments were received regarding grammatical wording within the DERP/EA.
Response: Suggested changes were incorporated into the ERP/EA, where appropriate.

Comment: We would like the administrative record to show that the April 20, 2011 Framework for Early Restoration Addressing Injuries Resulting from the Deepwater Horizon Spill Agreement, between the Trustees and BP, which describes the early restoration arrangement and the money set aside to fund early restoration projects, has implementing criteria which are different and separate from the Trustee Allocation Agreement. The Trustee Allocation Agreement is an agreement solely among the Trustees, and BP is not bound by its terms.
Response: The Framework Agreement between the Trustees and BP and the Trustees' Allocation Agreement are both included in the Administrative Record, and their terms speak for themselves. The Trustees acknowledge that the Allocation Agreement is an agreement among the Trustees only.

Comment: One comment opposed the Orange Beach, Baldwin County area portion of a boat ramp project. Please remove this project and place this land in permanent "off limits" status.
Response: The DERP/EA does not include a boat ramp project in Baldwin County, Alabama.

Comment: The issue of carbon credits has not been a part of the DWH NRDA restoration discussion.
Response: The Trustees are unclear on the comment regarding the relationship between the goals of the NRDA process and carbon credits. The Trustees are unable to respond further.

Comment: A number of comments were submitted regarding matters such as general information or comments about the BP claims process, human health concerns, and spill response activities.
Response: The Trustees acknowledge these general comments are related to the Spill; however these are outside the scope of the NRDA Early Restoration process.

Comment: Provide more information regarding development, the need, and the process for calculating NRD offsets.
Response: The Trustees believe the level of detail already provided in the DERP/EA is consistent with the Framework Agreement, applicable laws, regulations and Pre-Trial Orders.

Comment: Provide an approximated statistical range (error bars) for proposed offsets.
Response: Proposed Offsets were negotiated with BP and fairly and reasonably reflect the estimated benefits of each project and include consideration of uncertainties.

Comment: Several submitters commented on potential injuries caused by the spill and spill response activities.
Response: The assessment of the injuries and losses to resources and services caused by the Spill, including those resulting from the response, is ongoing.

Comment: The Trustees should prepare and publish "statistic material" describing the main aspects of the spill and research problems connected with the spill's aftermath to evaluate proposed projects and further steps.
Response: The natural resource damage assessment process is ongoing and while the Trustees have released data and information as it becomes available in final form, many studies are not yet complete. The Trustees will continue to make information and analyses gained or developed as part of the assessment available.

Comment: People around the world are watching the NRDA restoration process and will hold the Trustees accountable.
Response: The Trustees are doing the best job they can to assess the natural resource injuries and compensate the public.

Comment: One commenter expressed love for and a willingness to fight to protect the natural resources of the Gulf of Mexico.
Response: The Trustees share this sentiment.

Comment: One comment asked for clarification of what "discounting" means in regards to offsets.
Response: Discounting adjusts for differences in the timing of project benefits, enabling calculation of the "present value" of a stream of future benefits and expression of offsets in comparable units. Discounting is commonly applied in the natural resource damage context.

5.2   Comments Specific to Proposed Projects

5.2.1   Lake Hermitage Marsh Creation – NRDA Early Restoration Project

Comment: Comments supported the Lake Hermitage Marsh Creation – NRDA Early Restoration Project as a DWH NRDA restoration activity.
Response: The Trustees acknowledge support for this proposal.

Comment: Consider a collaborative partnership with the Restore the Earth Foundation and incorporating the use of Gulf Saver bags into the Lake Hermitage Marsh Creation – NRDA Early Restoration Project design.

Response: The contract to conduct vegetative plantings at the Lake Hermitage Marsh Creation – NRDA Early Restoration Project site will be advertised on the Louisiana Procurement and Contract Network website (http://wwwprd.doa.louisiana.gov/osp/lapac/pubmain.asp). Any entities that wish to be eligible bidders must register as a vendor with the State of Louisiana (http://www.doa.louisiana.gov/osp/vendor_index.htm).

Comment: Project cost is $13,200,000 for 104 acres, or $127 thousand per acre. The cost of projects in the recently released Louisiana's Comprehensive Master Plan for a Sustainable Coast is in this ball park, but there are other methods for restoration that are < 1/10th the cost per acre.

Response: As discussed in Section 3.2.2.1.2, the Trustees consider the estimated cost of this project to be reasonable. The estimated cost to carry out the project was one of many criteria considered by the Trustees in the selection of restoration projects. The DERP/EA provides substantial information about the other project selection criteria and the project selection process for Phase I early restoration.

Comment: Provide information supporting the cost estimate, project life span, and a return on investment analysis (e.g., if the marsh restoration subsides and dieback occurs 5, 10, 15 years following completion, what is the actual cost:benefit estimate).

Response: The initial cost estimate and estimate of project life span was based on engineering work completed by Louisiana's Coastal Protection and Restoration Authority for the larger CWPRRA-funded Lake Hermitage Marsh Creation project (BA-42). The project design life is 20 years. The primary design criterion is to obtain a marsh elevation of at least +1.20 ft. NAVD 88 throughout the life of the project. To accomplish this, the fill site template design includes an initial fill elevation of +2.0 ft. NAVD 88. According to engineering calculations, this should account for long term settlement and consolidation allowing the newly created marsh to meet the criteria of +1.20 ft. NAVD 88. While there is always uncertainty associated with restoration work, the Trustees believe that this project has a high likelihood of success.

Comment: What would be the learning/scalable lessons from this project which could be applied to further restoration at later stages of the NRDA process? I realize there is a perceived or real need for early restoration, but some of the activities seem disconnected and there is an opportunity/scale cost associated with a lack of a unified framework.

Response: The Lake Hermitage Marsh Creation – NRDA Early Restoration Project is consistent with Louisiana's 2012 Comprehensive Master Plan for a Sustainable Coast. The Trustees expect that the projects proposed in any future early restoration plan(s) and in any subsequent NRDA restoration plan(s) will provide a larger framework for coastal restoration actions. The Trustees will monitor the project and lessons learned from this monitoring will be used to improve future restoration efforts.

Comment: How will emergent vegetation dynamics be supported while the "restored" habitat rests in an ecological setting where no sediment inputs offset the background shallow subsidence and/or storm impacts?

Response: The project is in the outfall area of the West Point a la Hache Siphons, which will provide nutrients and suspended sediment via the diverted Mississippi River water, adding to the sustainability of the newly created marsh. Engineering and design have estimated subsidence and background erosion rates from historical storm activity. This project is expected to create marsh habitat and provide associated services over the project lifespan.

Comment: Would hydrologic exchange be impeded by the "earthen terraces"? Will there be impoundments of some kind on part or all of this project?

Response: The Lake Hermitage Marsh Creation – NRDA Early Restoration Project area will be constructed in place of the earthen terraces that were part of the original CWPPRA design for Lake Hermitage. Thus, there is no concern about hydrologic exchange being impeded by the "earthen terraces." There are no impoundments around this portion of the project.

Comment: I am concerned about residual DWH oil and dispersants. Sampling and analysis of sediment and marsh flora and fauna in the area must be performed to determine the extent and impacts of DWH oil and dispersants in the area prior to initiation of the project.

Response: The assessment is ongoing. The Trustees are still assessing the extent to which exposure of natural resources to oil and/or dispersants may have caused or contributed to any injuries or losses of natural resources or services. If, prior to or during implementation processes, issues arise regarding the contamination of the project area, the Trustees may utilize contingency funds to adaptively manage the problem.

Comment: DERP/EA description of the success monitoring and adaptive management plan for the project is insufficient.

Response: Language in section 3.2.2.1.3 ("Performance criteria, monitoring, and maintenance") regarding the Lake Hermitage Marsh Creation – NRDA Early Restoration Project was revised to add more specificity.

Comment: A lack of discussion exists regarding uncertainty for expected project outcomes, other than "expert opinion".

Response: The likelihood of success of each project is one of many criteria considered by the Trustees in the selection of restoration projects. For this first phase of early restoration, the Trustees focused on types of projects with which they have significant experience, allowing them to predict costs and likely success with a relatively high degree of confidence. The proposed Lake Hermitage Marsh Creation – NRDA Early Restoration Project is technically feasible and utilizes proven techniques with established methods and documented results. Local, state and federal agencies have successfully implemented similar marsh creation projects in the proposed region.

Comment: Every effort should be made to utilize the local workforce and create jobs through restoration.

Response: The Trustees expect that funding spent on restoration will create jobs and economic opportunity. For example, a recent economic analysis of marsh restoration at the Central Wetlands Unit (CWU) near downtown New Orleans estimated that the $72-million project is "on track to create 280 direct jobs and 400 indirect and induced jobs, for a total of 680 jobs over the project's life" (see http://www.estuaries.org/images/81103-RAE_17_FINAL_web.pdf). The

Trustees are required to abide by State procurement procedures for all contracts. The Trustees encourage local businesses to register as vendors with the State of Louisiana and reply to contracting opportunities as appropriate.

5.2.2   Louisiana Oyster Cultch Project

Comment: Comments supported the Louisiana Oyster Cultch Project as a DWH NRDA restoration activity.
Response: The Trustees acknowledge support for this proposal.

Comment: Designed artificial reef modules should also be considered in all the gulf oyster restoration projects due to their cost effectiveness and ability to generate more local jobs than oyster shells or crushed limestone.
Response: The proposed project involves placement of cultch materials onto public oyster seed grounds consistent with cultch planting approaches historically utilized by the Louisiana Department of Wildlife and Fisheries. This traditional approach typically supports many jobs including tug boat operators and deck-hands, spray barge deckhands, crane operators, and project managers. The contracts for cultch placement will be advertised on the Louisiana Procurement and Contract Network website (http://wwwprd.doa.louisiana.gov/osp/lapac/pubmain.asp). Any entities that wish to be eligible bidders must register as a vendor with the State of Louisiana (http://www.doa.louisiana.gov/osp/vendor_index.htm). The Trustees will continue to evaluate additional project ideas – including designed artificial reef modules – for funding as part of the NRDA restoration process and will continue to accept restoration project ideas at www.losco-dwh.com.

Comment: The Trustees should demonstrate how this project fits into a larger ecosystem restoration plan for the overall populations of Louisiana oysters.
Response: The proposed Louisiana Oyster Cultch Project has a specific objective to produce seed-sized and sack-sized oysters on public oyster seed grounds. This project has a clear nexus to resources injured by the Spill because oysters were exposed to oil, dispersant, as well as response activities undertaken to prevent, minimize, or remediate oiling from the Spill. This project is consistent with the Louisiana Regional Restoration Planning Program, the Louisiana Coastal Master Plan, the Louisiana Department of Wildlife and Fisheries' management of Louisiana's public oyster resources and the goal of compensating the public for natural resource injuries resulting from the Spill.

Comment: Comments provided potential design ideas.
Response: The Trustees note the commenter's design ideas and will provide these ideas to the Louisiana Department of Wildlife and Fisheries which is responsible for the design of oyster cultch projects on public oyster seed grounds in the State of Louisiana.

Comment: The oyster hatchery, production and cultch process should incorporate improved technology, I hope that research conducted at the new hatchery will encourage development of new techniques. Data and research from the new oyster facility as well as data and information from the oyster seed ground should be made publicly available to help facilitate development of a sustainable oyster production model.

Response: The Trustees agree with the importance of sharing data and information from the new oyster facility. The Louisiana Department of Wildlife and Fisheries has a history of conducting research on improved technology for oyster production and disseminating this information to the public. For example, a series of workshops in 2011 and 2012 are providing information on off-bottom culture techniques (http://www.wlf.louisiana.gov/news/33321). The new oyster facility will be available for University researchers and the Trustees expect that the facility will be an active location for data collection, research, and publication.

Comment: The Final Phase I Early Restoration Plan should provide more information regarding potential ancillary ecological benefits provided by the proposed oyster reef (cultch).
Response: Although the cultch projects may have ancillary ecological benefits, the Trustees proposed this project specifically for its benefits to produce seed-sized and sack-sized oysters on public oyster seed grounds to compensate for oysters injured by the Spill.

Comment: The proposed Louisiana oyster cultch areas must be evaluated to determine if negative impacts occur due to the dead (hypoxic) zone's low dissolved oxygen levels prior to implementation.
Response: Location selection criteria included a consideration of salinity regime and historical and recent oyster production, to avoid selecting locations where water quality prohibits oyster settlement and growth.

Comment: The proposed Louisiana oyster cultch areas must be evaluated to determine if negative impacts occur or could potentially occur due to fresh water diversion projects, and that areas that have been impacted or have the potential to be impacted by fresh water from diversion projects should not be selected for early restoration projects.
Response: Project locations were selected in consultation with the Louisiana Department of Wildlife and Fisheries and the Louisiana Governor's Oyster Advisory Committee. The project locations attempt to avoid conflicts with on-going or planned fresh water diversion projects expected during the lifespan of the cultch project.

Comment: Proposed oyster cultch areas should be evaluated for the presence/ongoing impacts from DWH oil and/or dispersants prior to implementing the project.
Response: Louisiana intends to evaluate proposed oyster cultch areas for the presence/ongoing impacts from DWH oil and/or dispersants prior to implementing the project. Site locations not safe for oyster production will not be used.

Comment: Oyster projects should be selected based on a regional oyster restoration strategy to help ensure strategic design and sustainability.
Response: The six (6) proposed cultch placements within this project span five (5) coastal parishes of Louisiana. Project locations were selected based on their expected ability to produce seed-sized and sack-sized oysters on appropriate public oyster seed grounds. The locations were thoroughly vetted based on review of salinity regime, coastal restoration, oil and gas infrastructure, water depth, presence of hard-bottom, and current and historical oyster production.

Comment: Provide clarification on whether remaining cultch areas along the Louisiana coast will be restored and able to be harvested.

Response: Management of cultch areas proposed in the DERP/EA is not intended to prohibit restoration or harvesting of other areas. Louisiana will continue with its regular cultch placement program in addition to developing early restoration projects.

Comment: The oyster cultch project could support several ecosystem services in Louisiana's estuarine environment provided the project is designed and sited appropriately. However, the DERP/EA did not make it clear whether proposed cultch placements would be designed to support ecosystem services or oyster fishery production.

Response: Although the cultch projects may have ancillary ecological benefits, the Trustees proposed this project specifically for its benefits to produce seed-sized and sack-sized oysters on public oyster seed grounds, compensating for oysters injured by the Spill.

Comment: The description of the success monitoring and adaptive management strategy in the DERP/EA was insufficient; its augmentation should be similar to what the DERP/EA described regarding the oyster cultch proposal in Mississippi.

Response: Language in section 3.2.2.2.3 ("Performance criteria, monitoring, and maintenance") regarding the Louisiana Oyster Cultch Project was revised to add more specificity.

Comment: Evaluate other potential oyster hatcheries in Terrebonne Parish and the Western part of Louisiana.

Response: The proposed projects in the DERP/EA represent only the first phase of the early restoration process. The Trustees continue to evaluate additional projects for funding as part of the early restoration process. The Louisiana Trustees continue to accept restoration project submittals at www.losco-dwh.com.

Comment: Hire local work forces to implement the Louisiana Oyster Cultch Project.

Response: The contracts associated with this project will be advertised on the Louisiana Procurement and Contract Network website (http://wwwprd.doa.louisiana.gov/osp/lapac/pubmain.asp). Any entities that wish to be eligible bidders must register as a vendor with the State of Louisiana (http://www.doa.louisiana.gov/osp/vendor_index.htm). The Trustees are required to abide by procurement procedures, consistent with relevant regulations and policy. The Louisiana Trustees encourage local businesses to register as vendors with the State of Louisiana and reply to contracting opportunities as appropriate.

Comment: The EA for the Louisiana Oyster Cultch Project is sufficient.

Response: The Trustees acknowledge the support for an Environmental Assessment of this proposed project.

5.2.3 Mississippi Oyster Cultch Restoration

Comment: Comments provided support for the project.
Response: The Trustees acknowledge these comments.

Comment: Opposition was expressed to the project due to and/or concern regarding potential impacts of residual DWH oil.

Response: The Deepwater Horizon NRDA includes assessment and evaluation of potential injuries to a wide array of natural resources. The assessment of potential impacts to shellfish and other sensitive resources is ongoing. Cultch placement areas would be screened prior to project implementation. Site locations not safe for oyster production will not be used.

Comment: An EA is sufficient for evaluating impacts from this project.

Response: The Trustees acknowledge the support for an EA of this proposed project.

Comment: Hire a local workforce to implement this project.

Response: The Trustees will abide by procurement laws, regulations and policy. These may vary, depending on the agency implementing a project. To the extent permissible by law, the Trustees will give preference to local hires.

Comment: Comments proposed modifications or questions regarding project design, such as: add additional oyster restoration areas and/or spawning beds to the design; revisit the design of the proposal, it will not effectively restore damaged oyster reefs; conduct a pilot project; add an oyster farming or relay component; dredge oyster beds prior to deploying cultch; modify timing and location of cultch deployment; "spark" the existing oyster reefs to evaluate the potential for natural recovery prior to expending funds deploying cultch; designed artificial reef modules should be considered in all the gulf oyster restoration projects due to their cost effectiveness and using artificial reef modules in order to generate more local jobs than oyster shells or crushed limestone; why have you chosen to seed oysters on the water bottoms rather than using another method of growing oysters suspended?

Response: The state of Mississippi has approximately 12,000 acres of total cultch areas. Cultch placement areas have been identified and permitted for cultch deployment by the Mississippi Department of Marine Resources as part of its on-going program. This early restoration project focused on areas of planned and permitted oyster cultch placement that could be completed in a timely manner. Oyster seeding has been used for oyster production in Mississippi, but was not a part of the project proposal. Designed artificial reef modules are not a part of the Mississippi Oyster Cultch Restoration Project. Traditional cultch placements typically support many jobs including tug boat operators and deck-hands, spray barge deckhands, crane operators, and project managers.

The Trustees note the commenter's design ideas and will provide these ideas to the Mississippi Department of Marine Resources which is responsible for the design of oyster cultch projects on public oyster harvest areas in the State of Mississippi. New project ideas such as "sparking", relay (oyster seeding), oyster farming, and the use of artificial reef modules in oyster restoration may be submitted at: http://www.mdeqnrda.com/ or http://www.gulfspillrestoration.noaa.gov/restoration/give-us-your-ideas/suggest-a-restoration-project/.

Comment: Clarify whether remaining cultch areas in the Mississippi Sound will be restored and able to be harvested.

Response: A total of 1,430 acres is proposed for the early restoration Mississippi oyster cultch placement. Cultch placement areas have been identified and permitted for cultch deployment based upon existing cultch programs in the Mississippi Sound. Location selection criteria included consideration of salinity, historical and recent oyster production, and substrate.

Comment: This proposal does not address the potential incompatibility of freshwater diversions with oyster survival or the conflicts that may arise among different restoration goals.
Response: Oyster cultch placements will be deployed in areas where existing and historical oyster reefs are present. Project locations will be selected to avoid conflicts with on-going or planned freshwater diversion projects expected during the lifespan of the cultch restoration project.

Comment: The Mississippi Oyster Cultch Restoration project could support several ecosystem services in Mississippi's estuarine environment provided the project is designed and sited appropriately. However, the DERP/EA did not make it clear whether proposed cultch placements would be designed to support ecosystem services or oyster fishery production.
Response: Although the cultch projects may have ancillary ecological benefits, the Trustees chose this project specifically for its benefits to increased secondary production in the form of harvestable oysters in the Mississippi sound. Oysters produced could ultimately be harvested by oystermen.

Comment: The description of the success monitoring plan was articulated better than others in the DERP/EA.
Response: The Trustees acknowledge this comment.

Comment: Description of the project in the DERP/EA failed to provide descriptions of the timing of short and long-term assessments of project success, the actual numeric metrics by which project success would be measured, what conditions would trigger corrective actions, and a description of corrective actions.
Response: The monitoring program would determine whether the project goals and objectives have been achieved, or whether corrective actions are required to meet the goals and objectives. Project performance may be assessed through physical and biological monitoring of oyster cultch plants. Mid-course enhancements may include additional cultch placement in areas of cultch loss or failed spat set.

Comment: Restore water quality in oyster bed areas to facilitate oyster restoration/recovery.
Response: New project ideas such as restoring water quality in oyster bed areas may be submitted at: http://www.mdeqnrda.com/ or http://www.gulfspillrestoration.noaa.gov/restoration/give-us-your-ideas/suggest-a-restoration-project/.

Comment: Figure 9 ("Oyster production (in Sacks of Oysters harvested), 2008 to 2011") is misleading: oyster harvest in 2011 would have been negatively affected from the exceptionally high river discharge in 2011.
Response: The sources for these seasonal oyster harvest values are as follows: 2008-2009 from Mississippi Department of Marine Resources 2009 Comprehensive Annual Report, Fiscal Year

Ended June 30, 2009; 2009-2010 from Mississippi Department of Marine Resources 2010 Comprehensive Annual Report, Fiscal Year Ended June 30, 2010; and 2010-2011 from Mississippi Department of Marine Resources, Coastal Markers, Volume 14, Issue 4, Summer 2011. The 2011-2012 oyster harvest values were not yet available as this ERP/EA was being published.

Comment: Develop a citizen's advisory group to provide input into restoring the local oyster fishery.

Response: There currently is an existing Oyster Task Force in Mississippi that meets with regulatory agencies approximately four times a year. The Task Force consists of representatives of dredgers, tongers, recreational fishermen, processors, and the Vietnamese community. One of the Task Force's goals is to make suggestions to regulatory agencies pertaining to management of Mississippi's oyster fisheries.

Comment: Any plans by state agencies and/or their engineering consultants to spend excessive fees and/or to purchase additional equipment, vehicles, vessels, computers, etc., with the restoration funds that should be spent to restore Mississippi's oyster resources should be stopped immediately.

Response: The funds would be expended to restore and enhance approximately 1,430 acres of the oyster cultch areas within the Mississippi Sound. Cultch material (oyster shell, limestone or crushed concrete, or some combination thereof) would typically be deployed at a rate of 100 cubic yards per acre with adjustments for site conditions as needed.

5.2.4 Mississippi Artificial Reef Habitat

Comment: General comments provided support for this proposal.

Response: The Trustees acknowledge support for this proposal.

Comment: One Commenter opposes artificial reefs due to the fact that they can serve as an attraction for fishing and can lead to over-exploitation of resources that were already decimated by the Spill.

Response: The phenomena of attraction versus production of fish (tertiary producers) associated with created reefs sites is debated in the literature. The extent to which existing artificial reef sites in Mississippi are contributing to exploitation of resources is occurring is unknown. Secondary production is nonetheless very important to the overall productivity of marine environments and Mississippi's nearshore artificial reef project was designed to enhance secondary producer trophic levels in waters where that productivity was injured.

Comment: One Commenter opposes the proposal because it will introduce a hard bottom structure that has the possibility of being invaded by invasive species.

Response: Hard bottom habitats are currently present in the Mississippi Sound in the form of the existing nearshore artificial reefs and adjacent hard bottom habitats. These artificial reefs and associated hard bottom habitats are permitted in the existing areas with USACE Nationwide Permit SAM-2011-01777-SPG. This proposal would enhance those existing hard bottom habitats as material would only be deployed within the current reef footprint.

Monitoring will be performed following implementation and any necessary maintenance or management activities would be instituted which would help mitigate and/or quantify the risk and possibly provide an opportunity for early control of any invasive species. The net benefit of cultch placements in nearshore reefs is anticipated to be positive.

Comment: Hire a local work force to implement this project.
Response: The Trustees will abide by procurement laws, regulations and policy. These may vary, depending on the agency implementing a project. To the extent permissible by law, the Trustees will give preference to local hires.

Comment: Reevaluate the project design, including evaluating the potential of an oyster reef in this area in lieu of an artificial reef.
Response: The artificial reefs are low profile patch reefs used to promote secondary productivity in the nearshore environment. The low profile reefs are constructed in a largely similar manner to the oyster cultch projects, but to a slightly higher elevation (4-6" as limited by permit), and will provide benefits comparable to the proposed oyster cultch reefs. The Mississippi early restoration projects for oyster cultch and artificial reefs are focused on secondary productivity.

Comment: I question the likelihood of success (e.g., whether the material will simply subside and/or be moved by storms) and ask for more information regarding project success monitoring and adaptive management (provide information regarding what conditions would trigger corrective actions and a description of what the corrective actions would entail).
Response: The nature of the proposed low-profile nearshore reefs, patches of cultch material for encrusting growth, is more likely to get buried rather than be transported any significant distance. In addition, natural spat settlement and colonization of the cultch material will stabilize the material within 12 to 18 months immobilizing a large fraction of the materials. Monitoring of existing nearshore reefs by the MDMR demonstrate that materials have not moved with previous storms. Colonization by various encrusting organisms is underway in a matter of weeks in successful cultch placements, which helps immobilize the materials.

Physical monitoring and biological monitoring of the reef will be implemented to assess the structural and biological integrity of nearshore reefs. Findings from the monitoring would be used to determine reef success, as well as maintenance and management activities. Maintenance, management, and corrective activities could include replacement of cultch degraded by environmental conditions.

Comment: Deployment of material through implementation of this project would negatively affect (e.g., bury) benthic communities already in the area.
Response: Cultch placement sites will be screened prior to cultch deployment. Cultch placement will be limited to areas that are existing hard bottom substrate or existing artificial reef areas. These artificial reefs and hard bottom habitats are permitted within the currently existing areas with USACE Nationwide Permit SAM-2011-01777-SPG. During the implementation of this project all effort would be made to avoid existing environmentally sensitive areas including any existing benthic communities. The net benefit of enhanced secondary productivity is anticipated to outweigh adverse effects of incidental filling and minor impacts to existing benthic communities.

Comment: Potential effects of residual DWH oil and/or other contaminants of concern could negatively impact the project.

Response: Mississippi intends to evaluate proposed nearshore reef cultch areas for the presence/ongoing impacts from DWH oil and/or dispersants prior to implementing the project. Site locations that are not safe for nearshore reef placement will not be used. If such issues arise prior to and during project implementation, the Trustees may be able to utilize contingency funds to modify project design, timing and/or otherwise adaptively manage problems.

Comment: Add permeable breakwaters to the artificial reef design.

Response: The Trustees considered this suggestion but do not feel it is necessary for achieving the project's purpose of expanding nearshore reefs. The current project is to expand existing nearshore reefs by placing cultch material in adjacent permitted areas. The activity is permitted under a Nationwide Permit which allows maintenance of cultch material on existing hard substrate. In the nearshore environment, navigation issues could be a concern with the proposed design and may require a modification to the U.S.ACE permit. Breakwaters are typically not a component of the artificial reef enhancements in shallow nearshore environment.

Comment: The proposed project has a poor nexus to restoring lost secondary productivity from the DWHOS and a sub-standard offset estimate as a measure of success. Therefore, this project should not be funded as an early restoration project. The Trustees did not clearly define the shallow water resources that would be restored by the expansion of artificial reef habitats, and how this would compensate for losses in secondary production from benthic habitats in Mississippi Sound. Estimating total offsets from these projects by improving biomass production would be difficult, especially if the project area is used by recreational fisheries.

Response: While there are comprehensive, ongoing efforts to assess natural resource injuries from the Spill, it is clear that one area of injury will be to the secondary producer trophic level. The proposed artificial reefs are low profile reefs (4-6" as limited by permit) in the nearshore environment that would be implemented to promote and to restore lost secondary productivity in the nearshore environment. The creation of nearshore reefs will enhance the recovery of various secondary productivity of invertebrate infaunal and epifaunal biomass at nearshore reefs. This enhancement of secondary producers will not expand artificial reef footprints but simply work within the current footprint permitted by USACE Nationwide Permit SAM-2011-01777-SPG. Offsets reflect estimated kilograms of biomass produced, and would be applied against secondary productivity injuries in the Mississippi Sound from the Spill as determined by the Trustees' total assessment of injury.

Comment: The Mississippi Artificial Reef Habitat proposal is more appropriate as a human use project than an ecological benefit project.

Response: The Trustee designed the Mississippi Artificial Reef Habitat project for the benefit of secondary production of invertebrate infaunal and epifaunal biomass at nearshore artificial reefs.

Comment: An EIS should be conducted for this project due to potential collateral injury to other natural resources or services.

Response: Evaluated on its own, the project would not have a significant environmental impact for the following reasons: 1) The permitted work is limited to deployment of cultch on existing reefs which mainly consist of hard reef substrate of limestone or concrete as well as a very limited amount of soft sediments of sand, silt, or clay. Cultch placement would be restricted to these existing hard substrates with only a limited amount of soft sediments cultched as an incidental impact. 2) Cultch replenishment on existing reefs will occur in small areas that are widely dispersed. A total of 100.5 acres of cultch would be deployed over a total of 67 sites. The project was authorized by the U.S. Army Corps of Engineers (USACE) under Nationwide Permit 4-Fish and Wildlife Harvesting Enhancement and Attraction Devices and Activities. The USACE typically uses Nationwide Permits for low impact, routine activities and completes an environmental review of these activities. 3) An endangered species act assessment and consultation on the project resulted in concurrence that project activities were not likely to adversely affect endangered or threatened species. 4) Essential Fish Habitat consultation with the National Marine Fisheries Service (NMFS) has been completed as part of the ERP/EA, which concluded that the project would not adversely impact EFH. 5) A complete review of this project under Section 106 of the Historic Preservation Act will be completed prior to project implementation.

For the reasons stated above, we believe that an EA is sufficient for the Mississippi Artificial Reef Habitat Project.

Comment: Natural events such as hurricanes could cause the artificial reef to shift, potentially scouring the sea bottom, migrating into sensitive marsh or sea grass habitat and counteracting the ecological and human use benefits.
Response: The nature of the proposed low-profile nearshore reefs, patches of cultch material for encrusting growth occurs quickly therein ensuring stabilization of the reef to withstand storm events. In addition, natural spat settlement and colonization of the cultch material will stabilize the material within 12 to 18 months immobilizing a large fraction of the materials. Monitoring of existing nearshore reefs by the MDMR demonstrate that materials have not moved with previous storms. Colonization by various encrusting organisms is underway in a matter of weeks in successful cultch placements, and help immobilize the materials.

Physical monitoring and biological monitoring of the reef will be implemented to assess the structural and biological integrity of nearshore reefs. Findings from the monitoring would be used to determine reef success, as well as maintenance and management activities. Maintenance, management, and corrective activities could include replacement of cultch degraded by environmental conditions.

Comment: Limestone has a low level of toxicity; however, the material should be tested prior to deployment to determine chemical composition.
Response: The MDMR Artificial Reef Bureau has constructed, managed, and monitored artificial reef areas for many years within Mississippi waters. This program follows all necessary best management practices as well as federal and state statues to protect natural resources. Cultch material is typical oyster shells, clam shells, limestone or crushed concrete. Cultch material must be clean and free of any hazardous substances.

Comment: Repeated deposits of crushed limestone may be necessary to counteract sedimentation and maintain artificial reef functionality over time, raising concerns about additive environment impacts of subsequent applications of introduced substrate.

Response: Artificial reef material deployments do not have an infinite lifespan. It is understood that additional applications of artificial reef material may be necessary for maintenance and/or management activities throughout this time period. In addition, the project was authorized by the U.S. Army Corps of Engineers (USACE) under Nationwide Permit (SAM-2011-01777-SPG) which is typically used for low impact, routine activities for which an environmental review has been performed.

Comment: Artificial reefs can damage nets trawled by shrimp fishing boats and reduce the available area for shrimp trawling.

Response: In general, nearshore reef locations are in waters too shallow for shrimp trawling. There are 67 existing nearshore artificial reefs in Mississippi waters which are managed by MDMR's Artificial Reef Bureau. The artificial reef deployment will only take place within the current reef footprints of the 67 existing nearshore artificial reefs.

Comment: Significant environmental impacts, risks, and uncertainties are not adequately considered in the Draft ERP and EA.

Response: The artificial reef project was authorized by the U.S. Army Corps of Engineers (USACE) under Nationwide Permit (SAM-2011-01777-SPG) which is typically used for low impact, routine activities for which an environmental review has been performed. Endangered Species Act (ESA) and essential fish habitat (EFH) consultation with the National Marine Fisheries Service (NMFS) has been completed on the ERP/EA. In addition, the Mississippi Department of Marine Resources has issued a certification in compliance with the Mississippi Coastal Wetlands Protection Act. Miss. Code Ann. § 49-27-1, et seq. This certification also serves as the coastal zone consistency certification for the purposes of the Coastal Zone Management Act and the Mississippi Coastal Program (DMR-120097; October 28, 2011).

Comment: Due to its precedential nature, potential for cumulative impacts on the environment, and the controversial nature of artificial reefs in the scientific community, the project will have a "significant" effect on the environment.

Response: Since April 2007, the MDMR Artificial Reef Bureau has conducted 141 inshore material deployments in the three coastal counties. Nearshore artificial reef material deployment will be completed on the 67 existing nearshore artificial reefs or hard substrate habitats although a limited amount of soft bottom habitat could be incidentally impacted. The 100.5 acres of material will be placed throughout the existing 67 sites and only within the current reef footprint. The project was authorized by the U.S. Army Corps of Engineers (USACE) under Nationwide Permit (SAM-2011-01777-SPG) which is typically used for low impact, routine activities for which an environmental review has been performed. As part of the DERP Endangered Species Act (ESA) and essential fish habitat (EFH) consultation with the National Marine Fisheries Service (NMFS) has been completed. Lastly, the proposed artificial reefs are low profile reefs (4-6 inches as limited by permit) in the nearshore environment and focus on enhancing secondary producer trophic levels. The construction of these artificial reefs would be similar to that for the oyster cultch project and would provide similar benefits in terms of secondary productivity.

5.2.5   Marsh Island (Portersville Bay) Marsh Creation

Comment: The EA for the Marsh Island proposal needs to be revised due to an inconsistency: page 85 states "Mississippi diamond-backed terrapins are a species of special concern and are known to exist in the project areas," and impacts would be temporary, but then states that "no adverse impacts to terrapins are expected."

Response: Any possible impacts to Mississippi diamond-backed terrapins are expected to be minor and temporary if they occur at all. Therefore, no significant or long-term adverse impacts are expected. The language on page 85 of the Marsh Island DERP/EA has been revised accordingly.

Comment: Comments objected to the Marsh Island restoration proposal and questioned why the Trustees did not wait till after Feb 14, 2012 (the close of the public comment period) to make a decision about approving the project.

Response: The Trustees actively solicited public input through a variety of mechanisms for potential restoration projects prior to development of the Phase I DERP/EA. Consistent with the Framework Agreement, each Trustee prioritized project proposals based on established selection criteria prior to bringing them forth for Trustee group consideration and initial negotiation with BP on cost and Offsets. The DERP/EA represented the Trustees' proposal of an initial list of projects. The Trustees considered all public comments on the DERP/EA prior to finalizing the selection of the Phase I early restoration projects in the ERP/EA.

Comment: Comments suggested using porous (nearly continuous rather than segmented) submerged designed breakwaters, such as designed artificial reefs, that mimic oyster reefs/natural environmental structures rather than emergent segmented breakwaters. Potential contractors have experience in these types of restorations and submerged breakwaters in regards to erosion problems. Designed artificial reef modules should also be considered in all the gulf oyster restoration projects due to their cost effectiveness and ability to generate more local jobs than oyster shells or crushed limestone.

Response: The final design of the breakwater portion of the project will be determined during the engineering and design phase of the project. The State of Alabama will consider many different breakwater designs and materials which meet the wave attenuation requirements for this environment and which also allow for adequate hydrological and biological exchange. Other pertinent design parameters will also be considered.

Comment: The state of Alabama used their Little Bay project as an example of how to stabilize the south shoreline for the Marsh Island (Portersville Bay) Restoration Project. The Little Bay project is attached to the mainland and used large concrete emergent Wave Attenuation Devices (WADs). In contrast, Marsh Island is a secondary barrier island in the middle of Mississippi Sound. A subtidal or intertidal breakwater would more closely mimic the natural oyster reefs and fit better into the natural landscape. The Nature Conservancy, Dauphin Islands Sea Lab and University of South Alabama have demonstrated these techniques both along mainland shorelines and on secondary barrier islands, and have shown similar effectiveness to the WADs. I would request that low crested subtidal or intertidal breakwaters be considered in the design, rather than emergent structures to limit large, man-made designs in an area with primarily natural landscapes.

Response: The final design of the breakwater portion of the project will be determined during the engineering and design phase of the project. The State of Alabama will consider many different breakwater designs and materials which meet the wave attenuation requirements for this environment and which also allow for adequate hydrological and biological exchange. Other pertinent design parameters will also be considered.

Comment: Regarding performance monitoring: "Project performance would be assessed by comparing quantitative monitoring results to predetermined performance standards that define the minimum physical or structural conditions deemed to represent normal and acceptable growth and development (e.g., elevation and colonization of native emergent vegetation)."
Comment: This description is sufficiently vague, but offers promise of doing it right.
Response: Additional monitoring details will be finalized as part of the engineering and design phase of the project.

Comment: The DERP states that "[p]roject performance would be assessed by comparing quantitative monitoring results to predetermined performance standards," but fails to define what those predetermined performance standards are or specify monitoring costs. The timing and public availability of short- and long-term assessments of project success is not described. A description of potential corrective actions is also absent.
Response: Additional monitoring details will be finalized as part of the engineering and design phase of the project. Monitoring data will be made publicly available. If project performance issues arise, the Trustees may be able to utilize contingency funds to modify project design, timing and/or otherwise adaptively manage problems. The specific type and extent of potential corrective actions can vary and will depend on the nature, extent, cause and other characteristics of underlying performance issues.

Comment: The proposed Alabama Marsh Island Restoration Project in Porterville Bay should not reduce the acres of public access to oyster floor and/or other public/recreational fishing use.
Response: There are currently no oyster beds within the project area and therefore the project will not reduce public access to oyster resources. Other than during project construction, the project site will remain open to public recreation use.

Comment: An EA needs to be completed as noted in DERP/EA, an EIS is potentially warranted, and public review of the EA needs to be completed prior to implementation.
Response: An environmental review under NEPA will be completed upon finalization of the design and construction details.

Comment: Use Ecosystems wave barrier instead of rip-rap.
Response: The final design of the breakwater portion of the project will be determined during the engineering and design phase of the project. The State of Alabama will consider many different breakwater designs and materials which meet the wave attenuation requirements for this environment and which also allow for adequate hydrological and biological exchange. Other pertinent design parameters will also be considered.

Comment: The Trustees need to better explain how the Marsh Island proposal fits into the overall coastal restoration goals of the Gulf and how it will be able to withstand factors such as

135

predicted rates of sea level rise and erosion over the next 50-100 years. This project requires a robust scientific analysis of the sustainability and effectiveness of this restoration project within a long term Gulf restoration plan prior to funding.

Response: The Trustees anticipate that the proposed project will restore Marsh Island to its approximate 1950 acreage and that the breakwater component of the project will protect the island from further wave-induced erosion for decades. The Marsh Island project is located within the Gulf ecosystem and is therefore part of, and relevant to, Gulf Coast restoration. Protection of marsh habitat is consistent with existing Gulf-wide restoration planning efforts, such as Gulf Coast Ecosystem Restoration Task Force (GCERTF, 2011), Mabus (2010), Brown et al. (2011), and Peterson et al. (2011). The final design of the breakwater portion of the project will be determined during the engineering and design phase of the project. The EA for Marsh Island, and its public review, will be completed prior to project implementation

Comment: The Marsh Island project appears doable, but it is questionable if the marsh will last. The history of the site, as described, is one of an eroding island. The island will not last without continuous supplemental dredge and fill. This is, therefore, unsustainable and not "restoration".

Response: The Trustees anticipate that the proposed project will restore Marsh Island to its approximate 1950 acreage and that the breakwater component of the project will protect the island from further wave-induced erosion for decades; continuous supplemental dredge and fill is not anticipated.

Comment: A plan for project success monitoring needs to be provided prior to project selection, as outlined by OPA (15 C.F.R. Section § 990.55(b)(2)).

Response: Project performance would be assessed by comparing quantitative monitoring results to predetermined performance standards that define the minimum physical or structural conditions deemed to represent normal and acceptable growth and development (e.g., elevation and colonization of native emergent vegetation). The monitoring program for this project would use these standards to determine whether the project goals and objectives have been achieved, or whether corrective actions are required to meet the goals and objectives. Details concerning the performance measures and monitoring will be developed prior to implementation of the project.

Comment: A public hearing on the detailed final design/scope of work for the Marsh Island project should be held, including a public comment period, prior to the issuance of a formal Request for Proposal.

Response: The Marsh Island Project will be subject to the public comment process of the USACE Individual Permit Process and the ADEM CZM/WQ Certification process. There will therefore be additional opportunity for public comment.

Comment: Consider a collaborative partnership with the Restore the Earth Foundation and it's Gulf Saver bag initiative for the Marsh Island (Portersville Bay) projects.

Response: Bids for the construction of the Marsh Island project will be placed and evaluated in compliance with Alabama Bid and Procurement law.

Comment: The Marsh Island project is too small for the expense; the project could be funded by other sources.

Response: The cost estimates for the Marsh Island project are similar to estimates for other large-scale marsh restoration projects. The Marsh Island project addresses marsh injury and it is appropriate to fund this project through the DWH NRDA early restoration process.

5.2.6 Alabama Dune Restoration Cooperative Project

Comment: Consider a submerged breakwater as part of the Alabama dune restoration project.
Response: Since the project will occur entirely on Gulf-side beaches, breakwaters would not be appropriate adjacent to sea turtle nesting habitat.

Comment: Proposed projects in Alabama may disrupt federally listed threatened or endangered species or sea turtle nesting.
Response: An Intra-Service Endangered Species Act Section 7 Biological Evaluation has been conducted with the U.S. Fish and Wildlife Service (FWS) regarding potential effects to federally listed species (Alabama beach mouse and associated critical habitat, loggerhead sea turtle, Kemp's ridley sea turtle, piping plover and associated critical habitat, and snowy plover). Based on the timing, location, logistics and best management practice guidelines of the project, the FWS concluded that implementation of the project was not likely to adversely affect these species. In summary, regarding nesting sea turtles, the Trustees will: a) avoid nesting season when possible; b) work during daylight hours; c) remove equipment from beaches each night to avoid interference with nesting turtles; d) survey for nests each morning to avoid impacting nest sites; and e) provide buffers to any nests encountered.

Comment: Comments objected to the Alabama dune restoration proposal because "the beach has already been ruined by dredging", and "residual oil could potentially ruin the restoration".
Response: The Trustees believe that it is important to begin restoration of dune habitats affected by the spill, including areas where renourishment projects have been completed. The Trustees believe that the threat of *Deepwater Horizon* oil to dune plant survival in the project location is low.

Comment: One comment objected to the Alabama dune restoration proposal at Gulf Shores and questioned why the Trustees did not wait till after Feb 14, 2012 (the close of the public comment period) to make a decision about approving the project.
Response: Consistent with the Framework Agreement, each Trustee prioritized project proposals based on a number of selection criteria prior to bringing them forth for Trustee group consideration, negotiation with BP, and inclusion into the DERP/EA. However, the Trustees considered all comments on the DERP/EA prior to final selection of Phase I early restoration projects.

Comment: Dune injury in Baldwin county was response-related; restoration should be funded directly by BP and not from a settlement with the Trustees (i.e., the Framework Agreement) that BP gets credit for through Offsets.
Response: The Framework Agreement allows for early restoration projects to address areas with injuries or impacts associated with response activities. It is intended to accelerate meaningful restoration in the Gulf in advance of the completion of the natural resource damage assessment

process. The Trustees will take this comment into account appropriately in the Natural Resource Damage Assessment process.

Comment: Planting sea oats is a good idea but plants should be planted between swells of existing dunes rather than building new dunes with dredged sand. Use sargassum weed where planting.
Response: Exact planting locations have not yet been selected. However, they are expected to encompass a variety of areas, including sites adjacent to existing primary dunes. The choice of plant fertilizer is outlined in the project proposal description, which has been used successfully in past projects. Active dune building is not included in the project design.

Comment: Clarify the project's relevance to Gulf of Mexico restoration, duration and cost of monitoring, and the success reporting process.
Response: This project represents an important step in restoring dune habitats in Alabama that were injured as a result of the Spill. Monitoring and success of the project will be based upon survival of the installed dune plants and completion of the sand fencing and signage portions of the project.

Comment: Keep impacts to endangered sea oats in mind.
Response: Sea oats are not a federally listed endangered species. However BMPs will be used to reduce environmental impacts in general.

Comment: Incorporate the use of EKO Dune Save bags into the restoration design.
Response: After considering a variety of potential engineering designs and scope of the proposal, the dune restoration methods outlined in the DERP/EA were chosen to maximize the probability for success of the project based on similar past projects.

Comment: The descriptions of performance criteria, monitoring and maintenance for this project provides a reasonable model for briefly describing monitoring, performance criteria, and response actions in a manner that allows for evaluation of short-term project success. However, the description fails to provide criteria to determine what constitutes long-term success of the project or what corrective actions might entail after the initial 90-day assessment. While this project provides a better model of performance criteria, monitoring, and maintenance than is provided for other projects in the ERP, important details are lacking. The project monitoring period should be stated, and should be years, not months.
Response: Planting methods and plant species selected for this project have been employed for decades in dune restoration projects in Alabama and northwestern Florida. As a result, much is known about the critical threshold for success. The 90-day planting survival period was chosen as a success milestone because our experience suggests plants that survive to this point are well-established and will thrive without intervention. After 90 days, natural processes (e.g., storms) will ultimately determine the fate of the dune, just as with other naturally derived foredunes. However, even if the project is impacted by storms or other natural processes, dunes developed by the project will function naturally, providing propogative material for the establishment of new dunes.

Comment: The EA is sufficient and a categorical exclusion is appropriate.

Response: The Trustees acknowledge this suggestion.

5.2.7 Florida Boat Ramp Enhancement and Construction Project

Comment: General comments both supported and opposed the DERP/EA boat ramp proposal.
Response: The Trustees acknowledge these comments.

Comment: Comments opposed boat ramp construction and enhancement in lieu of providing funding to support natural resource enhancement and restoration.
Response: The goal of the NRDA regulations within the Oil Pollution Act is to make the environment and public whole for injuries to natural resources and associated services resulting from an oil spill incident. The Spill caused natural resource injury and the loss of ecological and human uses services. The Trustees will seek compensation for the public for both types of loss. Proposed projects in the DERP/EA represent only the first phase of the DWH NRDA restoration process. The Trustees will continue to evaluate additional projects for funding.

Comment: An EIS should be conducted for the Florida boat ramp construction and enhancement proposal.
Response: The Trustees have evaluated the environmental impacts of the Florida boat ramp construction and enhancement proposal in an environmental assessment, and have concluded the project will not result in significant impacts to the human environment that would necessitate the preparation of an EIS

Comment: The Trustees should evaluate predicted levels of fishing effort to evaluate whether additional boat ramps are warranted.
Response: New boat ramps are intended to enhance boating access in support of several recreational activities, including pleasure boating, diving, watersports, and fishing. It is anticipated that the number of additional fishing trips resulting from these new or enhanced ramps would not result in significant increases in pressure on fish stock or changes in fishing patterns. Any associated effects will be localized and likely minimal, and outweighed by the overall benefits to all forms of boat-based recreation.

Comment: Implementation of the boat ramp construction and enhancement project should include associated fishing use and angling success monitoring.
Response: Once completed, the boat ramps included in this project will be included in the sites sampled as part of NOAA's Marine Recreational Information Program.

Comment: Future recreation fisheries projects should include improvements to recreational fisheries monitoring.
Response: The Trustees will consider this input as DWH NRDA restoration proceeds.

Comment: The proposed boat ramp project may increase recreational traffic, increase recreational fishing and impact natural resources that were damaged, and should therefore not be funded.
Response: This project provides enhanced boating access in Escambia County, compensating for human loss of recreational access to the Gulf and surrounding waters. Natural resource and

environmental impacts were taken into consideration and were found to be minimal. Those potential secondary and cumulative impacts were balanced against the need to compensate for lost human use, and ultimately the Trustees determined this project to be appropriate for early restoration funding.

Comment: Concern was expressed that the human use injury would be emphasized (over ecological injuries) in Florida.
Response: Projects proposed in the DERP/EA represent only the first phase of the DWH NRDA restoration process. The Trustees continue to evaluate additional projects for funding as part of the early restoration process.

Comment: The boat ramp construction and enhancement proposal is not an appropriate NRDA restoration project because it does not restore natural resources.
Response: The goal of the NRDA regulations within the Oil Pollution Act is to make the environment and public whole for injuries to natural resources and associated services resulting from an oil spill incident. The DWH spill caused natural resource injury and the loss of ecological and human uses services. The Trustees will seek compensation for the public for both types of loss. Boat ramp construction and enhancement projects can be appropriate NRDA restoration projects, and have been utilized to address human use losses arising at other Oil Pollution Act incidents.

Comment: Law enforcement in the area of the proposed Perdido River boat ramp would be difficult due to location, mixed jurisdiction, and historic crime issues.
Response: Law enforcement routinely operates in the Perdido River and is accustomed to working in an area involving state line jurisdictional issues. FWC's law enforcement has a good working relationship with both the Alabama Marine Patrol on saltwater issues and the Alabama Fish and Game for hunting and freshwater issues. Although there are historic crime issues, FWC partnered with the Northwest Florida Water Management District to initiate an enforcement strategy to address them. Over the past three years, FWC documented successful enforcement efforts to reduce criminal behavior and return the recreational areas along the scenic Perdido River to a safe, family environment.

Comment: Boat ramps incorporating breakwaters would provide habitat for fish, crabs and shellfish, and offer boater protection from wave action.
Response: Oyster reefs or other breakwaters, as well as shore plantings, will be incorporated into boat ramp construction designs to provide protection and habitat. The two existing ramps discussed in the DERP/EA proposal already have wave attenuation devices attached to the piers adjunct to the ramp to provide protection.

Comment: Provide additional information on the selection of a boat ramp proposal; the inclusion of the boat ramp proposal is sufficiently different from the other projects to raise questions about the selection process.
Response: The Trustees utilized project selection criteria as identified in the ERP/EA. These criteria are consistent with applicable laws, regulations and the Framework Agreement. Phase I projects were identified through a reasonable balancing of early restoration project objectives,

opportunities and timelines in the process of applying project selection criteria. The Trustees believe the ERP/EA provides sufficient information regarding early restoration project selection.

Comment: How is the boat ramp construction and enhancement project a priority and establishing baseline data for marine mammal populations (for injury assessment and restoration) is not?
Response: Proposal of the boat ramp construction and enhancement project as part of the DWH early restoration process is consistent with goals of NRDA and the Framework Agreement, and does not displace any planned or potential studies of life history information related to marine mammals in the Gulf.

Comment: The description of the ramp use during the spill describes closure of the ramps, not damage to the ramps. There is no restoration, because the ramps can now be used again.
Response: Boat-based recreation was adversely impacted by the Spill, irrespective of potential physical damage to boat ramps. This project provides enhanced boating access in Escambia County, which will help offset boat-based recreational loss and is appropriate for early restoration.

Comment: I would strongly recommend that any approved project that expands access to coastal areas (in this case, construction of boat ramps) should include measures to protect sensitive shorebirds and seabirds from likely increases in disturbance. Examples of what this may involve include: 1) pre-posting historic shorebird/seabird nesting areas, 2) providing materials and transportation to volunteer bird stewards at accessible sites, and 3) dedicating funds from the project budget to pay for contract Law Enforcement officers to protect birds from disturbance. These measures would help prevent corresponding increases in disturbance to nesting and roosting shorebirds as public access to the area is facilitated.
Response: Potential impacts to sensitive habitats and fish and wildlife are an important consideration for all projects and are being addressed through the environmental analysis conducted for each project included in the Phase I ERP/EA. The boat ramps being constructed are on developed sites where it is not likely that nesting shore and seabirds will be impacted. Precautions will be taken in contracting for projects such as the boat ramp to ensure that the contractor is aware of regulations requiring protection of migratory birds and endangered species and that the appropriate wildlife permits are obtained if needed.

Comment: Implementing this project could result in alteration and/or damage to natural habitats, result in more collisions with marine animals, increase habitat loss, introduce chemicals (gas, oil) into water. Are these factors consistent with ecosystem restoration goals? Environmental impacts from construction, such as channel dredging or displaced species, were not discussed in proposal.
Response: The sites for the proposed two new boat ramp projects have been previously developed, with one of the proposed locations for a new boat ramp being a former industrial site and the other proposed location being a former single family home site with a covered boat slip in poor condition. There is little risk of additional alteration or loss of natural habitats at these sites. Florida's regulatory authorization, which is issued either in the form of permit or authorization letter from the Florida Department of Environmental Protection, is required for all the boat ramp projects. This authorization requires Best Management Practices (BMPs) for

erosion and sediment control to be utilized, which will help reduce damage and loss of habitat. Any dredging activities will be permitted by the same authorization and have the same BMPs requirement. For the three sites that have already been permitted, no submerged aquatic vegetation was observed at the sites; the Florida Department of Environmental Protection determined that fish and wildlife resources would most likely be unaffected. None of the proposed project sites are adjacent to manatee protection zones or sea turtle nesting habitat, so the risk of collision around the boat ramps is low. The US Army Corps of Engineers (USACE) authorization, which is issued either in the form of a 404 permit or an authorization letter pursuant to Section 10 of the Rivers and Harbors Act of 1899, is also required for all the boat ramp projects. Included in the USACE authorization are standard manatee conditions for in-water work as well as National Marine Fisheries Service sea turtle and smalltooth sawfish construction conditions, which ensure construction will not adversely affect those species. Any chemical releases are expected to be negligible and are not expected to result in significant impacts to natural resources.

Comment: Commenters submitted questions on cost (e.g., Most nice ramps are $500,000 at most). Does this include dredging? How will this money be tracked once it is given to Escambia County? There is confusion over the cost of the projects as listed in the public statements, versus the costs and benefits of the projects as listed by Florida Department of Environmental Protection. The Boat Ramp project is a merger of two such projects in Escambia County, but the final price exceeds the combined cost. Commenters expressed opposition to the project due to the estimated cost.
Response: Estimated costs of the boats ramps include all applicable costs, including design and implementation. Contingency and operation and monitoring costs, part of the proposed costs in the DERP/EA, were not included in the original proposal submitted to the Florida Department of Environmental Protection. Another factor in cost discrepancies is that the DERP/EA proposal includes road access improvements to the two proposed boat ramp sites which increased cost estimates. Financial oversight, auditing and reporting will follow applicable laws, regulations and policies.

Comment: Boat ramps are built and subsidized by a portion of fishing licenses, annual boat registrations, and annual boat trailer registrations, etc. as well as through the county. Why can't the county cover these costs through the bed taxes that are designed to support and enhance tourism?
Response: Construction and restoration of boat ramps helps compensate the public for the loss of boat-based access to natural resources due to the Spill. Escambia County proposed the boat ramp project for DWH NRDA early restoration funding because its recreational development needs exceed the available funding.

Comment: Provide a discussion regarding project success criteria. Omission of this information is unacceptable.
Response: Project success will be based on the boat ramps being certified upon completion that the ramps are built in general accordance with the plans, specifications and all specific permit conditions.

Comment: No Offsets are mentioned.

Response: The monetary Offset agreed to between the Trustees and BP ($10,153,642) is discussed in the Phase I ERP/EA, section 3.2.6.1.4.

5.2.8 Florida (Pensacola Beach) Dune Restoration

Comment: Comments were received in support of the project.
Response: The Trustees acknowledge this support.

Comment: The project should undergo a full Environmental Analysis, including a comprehensive description of the fertilizers used and the predicted timeline for watering and maintenance required to establish these plants. This project poses a significant impact and a categorical exclusion to NEPA review is not appropriate for a planting project of this scale.

Response: Because the dune restoration project involves planting and other minor revegetation actions, installing sand fencing and signage, and would result in only minor or negligible change in the use of the project area, U.S. Fish and Wildlife Service determined that the NEPA compliance would be covered as a Categorical Exclusion.

Comment: Additions should be made to the Florida dune restoration proposal, including: Navarre Beach; from the County land at the east end of the National Park Service's Fort Pickens (the Ft. Pickens Gate public park and beach areas) eastward to the end of Pensacola Beach's commercial area (the Marriott property), including the new Margaritaville Hotel and renovated Holiday Inn Express beach fronts; and an area adjacent to, just east of, the Holiday Inn Express on Pensacola Beach, 333 Ft. Pickens Road, and south of the hotel's parking lot. This area was used by BP and the Florida Dept. of Fish and Wildlife to access the beach. Numerous vehicles from 3 wheelers to large tractors crossed the dunes line here on a regular basis. This area was also used as a bus stop and drop off for BP workers (including the set of a port-a-potty). These workers walked across the dune line daily forming paths and trenches. Last fall someone poured a truck load of sand but this did not in any way restore the area back to its natural state.
Response: While the Trustees are not planning on expanding the Florida (Pensacola Beach) Dune Restoration project, the Trustees continue to evaluate additional projects for funding as part of the early restoration process.

Comment: The Florida dune restoration proposal should include raising and extending 24 dune walkovers in the project area.
Response: While the Trustees are not planning on expanding the Florida (Pensacola Beach) Dune Restoration project, the Trustees continue to evaluate additional projects for funding as part of the early restoration process.

Comment: Integrate EKO Dune Saver bags into the Florida dune restoration design.
Response: The project will be bid out for design and implementation. Specific project components will not be determined until the contract is awarded and the project design finalized.

Comment: Use equipment purchased (sand sharks, etc.) by Escambia County for the Santa Rosa Island Authority to be used along the beach and sound side areas to remove asphalt (dislodged from the road bed during previous hurricanes) before any form of replanting begins.

143

Response: The Trustees surveyed the site and do not believe that there is any asphalt in the project area that needs to be removed. However, if asphalt is encountered then it will be dealt with appropriately.

Comment: I am confused about the cost of the projects as listed in public statements versus the costs and benefits of the projects as listed by Florida Department of Environmental Protection. The Pensacola Dune project is only half funded, funded to 20 acres instead of the original 40 acres.

Response: The Pensacola Beach Dune Restoration, which is E-6 on the Florida Trustee's list of submitted projects, has always been 20 acres. The Pensacola Beach Renourishment, which is E-18 on the Florida Trustee's list, is listed as 8.2 miles or roughly 40 acres. This is a separate project that is being considered but not proposed for this phase.

Comment: What is the source and genetic quality of plants that will be used?

Response: All plants will be grown from seeds or cuttings from the Alabama coast or North Florida to ensure appropriate genetic stocks are used in the project.

Comment: Ongoing DWH spill response cleanup activities could potentially impact the project.

Response: Current clean-up in the area utilizes scoop nets in intertidal and supratidal areas but does not enter the dunes. Therefore, the Trustees believe that continuing response activities are unlikely to affect the success of this project. The Florida Department of Environmental Protection is overseeing this project as well as being involved in cleanup/response activities.

Comment: Articulate the relationship between this project and a longer-term restoration plan.

Response: The OPA regulations (15 CFR 990.54) include specific guidance on the utilization of existing restoration projects and regional restoration plans to address natural resource injuries when appropriate. Projects and strategies already developed under regional plans (e.g., the Gulf Coast Ecosystem Restoration Task Force (GCERTF, 2011), Mabus (2010), Brown et al. (2011), NRCS (2011) and Peterson et al. (2011) and restoration planning efforts undertaken by individual Trustees were considered in the early restoration process. The Trustees find this project to be consistent with existing and planned restoration efforts in the Gulf.

Comment: Provide performance criteria for determining what constitutes long-term success of the project, details regarding corrective actions after the initial 90-day assessment, and adaptive management plans to deal with erosion from storms or future re-oiling events and subsequent cleanup efforts.

Response: As indicated in the DERP/EA, project performance will be assessed by comparing quantitative monitoring results to predetermined performance standards that define the minimum physical or structural condition deemed to represent normal and acceptable growth and development. Additional monitoring details will be developed.

CHAPTER 6   LITERATURE CITED

Aten, L.E. 1983. Indians of the Upper Texas Coast. Academic Press, New York.

Bertness, M.D. 1999. The Ecology of Atlantic Shorelines. Sinauer Associates, Sunderland, MA.

Brown, C., Andrews, K., Brenner, J., Tunnell, J.W., Canfield, C., Dorsett, C., Driscoll, M., Johnson, E., and S. Kaderka. 2011. Strategy for restoring the Gulf of Mexico (a cooperative NGO report). The Nature Conservancy. Arlington, Virginia. 23 pp.

Coastal Environments, Inc. (CEI). 1982. Sedimentary studies of prehistoric archaeological sites. Prepared for the U.S. Dept. of the Interior, National Park Service, Division of State Plans and Grants, Baton Rouge, LA.

Christmas, J.Y. and R.S. Waller. 1973. Estuarine Vertebrates. In: Christmas, J.Y. (ed.). Cooperative GMEI. Phase IV, Biology. Gulf Coast Research Lab, p. 320-434.

Coastal Wetlands Planning, Protection, and Restoration Act (CWPPRA). 2011. Barataria Basin. World Wide Web electronic publication. http://lacoast.gov/new/About/Basin_data/ba/Default.aspx

Council on Environmental Quality (CEQ). 1997. Environmental Justice: Guidance under the National Environmental Health Policy Act. Washington, DC: President's Council on Environmental Quality.

Florida Fish and Wildlife Conservation Commission. 2009. Florida Boating Access Facilities Inventory and Economic Study Including a Pilot Study for Lee County. http://myfwc.com/about/overview/economic-benefits.

Gulf Coast Ecosystem Restoration Task Force (GCERTF). 2011. Gulf of Mexico regional ecosystem restoration strategy (preliminary). 104 pp.

Kirk, J.P. 2008. Gulf Sturgeon Movements In and Near the Mississippi River Gulf Outlet. U.S. Army Corps of Engineers, Engineer Research and Development Center. ERDC/EL TR-08-18.

Lewis, R.B. 2000. Sea-level Rise and Subsidence Effects on Gulf Coast Archaeological Site Distributions. American Antiquity 65(3): 525-541.

Louisiana Department of Wildlife and Fisheries (LDWF). 2008. L.D.W.F. Completes Oyster Reef Rehabilitation Project in Hackberry Bay. Press Release, 2008-175. http://www.wlf.louisiana.gov/news/29880 (Accessed April 6, 2012).

LDWF. 2009. Oyster Stock Assessment Report of the Public Oyster Areas in Louisiana: Seed Grounds and Seed Reservations. Oyster Data Report Series No. 15, July, 2010.

http://www.wlf.louisiana.gov/sites/default/files/pdf/page_fishing/32695-Oyster%20Program/2009_oyster_stock_assessment.pdf

LDWF. 2010. Oyster Stock Assessment Report of the Public Oyster Areas in Louisiana: Seed Grounds and Seed Reservations. Oyster Data Report Series No. 16, July, 2010. http://www.wlf.louisiana.gov/sites/default/files/pdf/page_fishing/32695-Oyster%20Program/2010-oyster-stock-assessment-report.pdf

Mabus, R. 2010. America's Gulf coast: a long term recovery plan after the Deepwater Horizon spill. 130 pp.

Mann, T., Mississippi Department of Wildlife, Fisheries and Parks. 2000, June 22. Letter to Susan Ivester Rees, Corps, Mobile District.

Minerals Management Service (MMS). 2007. Gulf of Mexico OCS Oil and Gas Lease Sales: 2007-2012 Final Environmental Impact Statement. U.S. Department of the Interior, Minerals Management Service, Gulf of Mexico OCS Region. http://www.boem.gov/Environmental-Stewardship/Environmental-Assessment/NEPA/nepaprocess.aspx.

Mississippi Department of Environmental Quality. 2010. 2010 Air Quality Data Summary. http://www.deq.state.ms.us/MDEQ.nsf/pdf/Air_2010AirQualityDataSummary/$File/2010%20Air%20Quality%20Data%20Summary.pdf?OpenElement.

Mississippi Department of Marine Resources (MDMR). 2010. Mississippi Department of Marine Resources 2009 Comprehensive Annual Report, Fiscal Year Ended June 30, 2009. http://www.dmr.ms.gov/images/dmr/2009-dmr-annual-report.pdf

Mississippi Department of Marine Resources (MDMR). 2011a. Mississippi Department of Marine Resources 2010 Comprehensive Annual Report, Fiscal Year Ended June 30, 2010. http://www.dmr.ms.gov/images/dmr/2010-dmr-annual-report.pdf

Mississippi Department of Marine Resources (MDMR). 2011b. Coastal Markers, Newsletter of the Mississippi Department of Marine Resources. Volume 14, Issue 4, Summer 2011.

Moretzsohn, F., J.A. Sánchez Chávez, and J.W. Tunnell, Jr., Editors. 2011. GulfBase: Resource Database for Gulf of Mexico Research. http://www.gulfbase.org, 17 July 2011.

National Oceanic and Atmospheric Administration (NOAA). 2011a. Hypoxia in the Gulf of Mexico: Progress towards the completion of an integrated assessment. Accessed September 23, 2011. http://oceanservice.noaa.gov/products/pubs_hypox.html.

Natural Resources Conservation Service (NRCS). 2011. Gulf of Mexico Initiative. U.S. Department of Agriculture. Washington, D.C.

NOAA. 2011b. The Gulf of Mexico at a glance: A second glance. http://stateofthecoast.noaa.gov/NOAAs_Gulf_of_Mexico_at_a_Glance_report.pdf.

NOAA, U.S. Department of the Interior, Louisiana Oil Spill Coordinator's Office, Louisiana Department of Environmental Quality, Louisiana Department of Natural Resources, Louisiana Department of Wildlife and Fisheries, 2007a. Regional Restoration Plan, Region 2, 24 pp. plus appendices.

NOAA, U.S. Department of the Interior, Louisiana Oil Spill Coordinator's Office, Louisiana Department of Environmental Quality, Louisiana Department of Natural Resources and Louisiana Department of Wildlife and Fisheries. 2007b, The Louisiana Regional Restoration Planning Program Final Programmatic Environmental Impact Statement, 172pp. plus appendices.

NOAA Fisheries. 2011a. Interactive Fisheries Economic Impacts Tool. NOAA Fisheries Economics & Social Sciences Program. http://www.st.nmfs.noaa.gov/st5/index.html.

NOAA Fisheries. 2011b. Monthly Commercial Landings Statistics Query. NOAA Fisheries Office of Science and Technology. http://www.st.nmfs.noaa.gov/st1/commercial/landings/monthly_landings.html.

Odum, W.E. 1988. Comparative Ecology of Tidal Freshwater and Salt Marshes. Annual Review of Ecology and Systematics 19:147-176.

Peterson, C.H., Coleman, F.C., Jackson, J.B.C., Turner R.E., Rowe, G.T., Barber, R.T., Bjorndal, K.A., Carney, R.S., Cowen, R.K., Hoekstra, J.M., Hollibaugh, J.T., Laska, S.B., Luettich, R.A., Osenberg, C.W., Roady, S.E., Senner, S., Teal, J.M., and P. Wang. 2011. A once and future Gulf of Mexico ecosystem: restoration recommendations of an expert working group. Pew Environmental Group, Washington, DC. 112 pp.

Pierce, M.E., and J.T. Conover. 1954. A study of the growth of oysters under different ecological conditions in Great Pond. Biol. Bull. (Woods Hole), v. 107, p. 318 (Abstract).

Posadas, Benedict C. 2001. Comparative Economic Analysis of Using Constructed Wetlands in Recirculating Catfish Pond Production. Journal of Applied Aquaculture, 11:1-20.

Posadas, B.C., and B.K.A. Posadas. 2011. Gulf Oil Spill Assessment Marine Sector Quarterly Report, Direct Employment Impacts of the Gulf of Mexico Oil Spill to Mississippi Seafood, and Commercial and Recreational Fishing Sectors. Mississippi State University, Coastal Research and Extension Center.

Priddy, R.R., R.M. Crisler, Jr., C.P. Sebren, J.D. Powell, and H. Burford. 1955. Sediments of Mississippi Sound and inshore waters. Miss. St. Geol. Surv. Bull. 82., University, Miss. 53p.

Ross, S.T., W.T. Slack, R.J. Heise, M.A. Dugo, H. Rogillio, B.R. Bowen, P. Mickle, and R.W. Heard. 2008. Estuarine and Coastal Habitat Use of Gulf Sturgeon (*Acipenser oxyrinchus desotoi*) in the North-Central Gulf of Mexico. http://www.fws.gov/filedownloads/ftp_panamacity/Pipl%20Literature/infauna/.

Thorpe, Paul, Ron Bartel, Patricia Ryan, Kari Albertson, Thomas Pratt and Duncan Cairns. 1997. The Pensacola Bay System Surface Water Improvement and Management Plan. Northwest Florida Water Management District, Publication.

U.S. Army Corps of Engineers. 2009. Mississippi Coastal Improvements Program, Hancock, Harrison, and Jackson Counties, Mississippi. Comprehensive Plan and Integrated Programmatic Environmental Impact Statement. Volume 1. Main report.

USACE. 2010. The U.S. waterway system: Transportation facts and information. Navigation Data Center. http://www.ndc.iwr.usace.army.mil/factcard/temp/factcard10.pdf

U.S. Department of Commerce, U.S. Census Bureau. 1999. Poverty. http://www.census.gov/hhes/www/poverty/.

USDA, NRCS. 2011. The PLANTS Database (http://plants.usda.gov, 13 December 2011). National Plant Data Team, Greensboro, NC 27401-4901 USA.

U.S. Energy Information Administration (USEIA). n.d. Gulf of Mexico fact sheet. http://www.eia.doe.gov/special/gulf_of_mexico/index.cfm.

USGS. 2002. Environmental Atlas of the Lake Pontchartrain Basin. USGS Open File Report 02-206. http://pubs.usgs.gov/of/2002/of02-206/.

U.S. Geological Survey and U.S. Environmental Protection Agency (USGS and EPA). 2011. ESRI maps, National Hydrography Dataset, EPA analyses. Courtesy of Stephen B. Hartley, USGS National Wetlands Research Center.

Vittor, B.A. and Associates. 1982. Benthic macroinfauna community characterizations in Mississippi Sound and adjacent waters. Final Report Contract No. DACW01-80-C-0427. Corps, Mobile District. 287p. plus appendices.

Weindorf, D.C. 2008. Land and Soil Resources of Louisiana, USA. Geographia Technica 2:85-108.

Appendix A

Deepwater Horizon Oil Spill Trustee Council Resolution Regarding: Approval of the Final Phase I Early Restoration Plan and Environmental Assessment for Release and Publication

(This page intentionally left blank.)

Trustee Council Resolution 12-1
ADOPTED *April 16, 2012*

**DEEPWATER HORIZON OIL SPILL TRUSTEE COUNCIL RESOLUTION
REGARDING:**

**Approval of the Final Phase I Early Restoration Plan and Environmental Assessment
for Release and Publication.**

1. In accordance with the Oil Pollution Act of 1990 (OPA), the National Environmental Policy Act (NEPA), and the Framework Agreement for Early Restoration Addressing Injuries Resulting from the *Deepwater Horizon* Oil Spill, the undersigned representatives of the Natural Resource Trustees hereby select the first eight early restoration projects as described in the Phase I Early Restoration Plan and Environmental Assessment (Phase I ERP/EA) and approve the release and publication of the Phase I ERP/EA to commence the process of restoring natural resources and services injured or lost as a result of the *Deepwater Horizon* oil spill, which occurred on or about April 20, 2010, in the Gulf of Mexico. The selected projects are:

    a. Lake Hermitage Marsh Creation- NRDA Early Restoration Project
    b. Louisiana Oyster Cultch Project
    c. Mississippi Oyster Cultch Restoration Project
    d. Mississippi Artificial Reef Habitat Project
    e. Marsh Island (Portersville Bay) Marsh Creation Project
    f. Alabama Dune Restoration Cooperative Project
    g. Florida Boat Ramp Enhancement and Construction Project
    h. Florida (Pensacola Beach) Dune Restoration Project

2. The Marsh Island (Portersville Bay) Marsh Creation Project is approved for completion of project design, NEPA analysis and work necessary to support applications for permits. NEPA review would be completed before any implementation occurs.

3. In selecting these projects and approving the Phase I ERP/EA, the Trustees are acting pursuant to the Oil Pollution Act of 1990 (33 U.S.C. § 2701 *et seq.*), the implementing Natural Resource Damage Assessment regulations found at 15 CFR Part 990, and the Framework Agreement for Addressing Injuries Resulting from the *Deepwater Horizon* Oil Spill.

4. This Resolution confirms and memorializes this approval for the Administrative Record.

SIGNATURES ON FOLLOWING PAGES:

Mimi Drew
Principal Representative for Florida Trustees

Trudy Fisher
Principal Representative for Mississippi

Deepwater Horizon Oil Spill Trustee Council Resolution 12-1
Page 4

Garret Graves
Principal Representative for Louisiana Trustees

Deepwater Horizon Oil Spill Trustee Council Resolution 12-1
Page 5

Carter Smith
Principal Representative for Texas Trustees

Craig O'Connor
Principal Representative for NOAA

Deepwater Horizon Oil Spill Trustee Council Resolution 12-1
Page 7

Cynthia Dohner
Cynthia Dohner
Principal Representative for the U.S. Department of the Interior

Cooper Shattuck
Principal Representative for Alabama Trustees

Appendix B

Federally Listed Threatened and Endangered Species with the Potential to Occur in Phase I Early Restoration Plan Proposed Project Areas

(This page intentionally left blank.)

Table B-1. Federally listed threatened and endangered species with the potential to occur in Phase I Early Restoration Plan proposed project areas.

Species	Threatened	Endangered
Wood Stork		X
Piping Plover	X	
Least Tern	X	
Alabama Sturgeon		X
Gulf Sturgeon	X	
Pallid Sturgeon		X
Smalltooth Sawfish		X
West Indian Manatee		X
Perdido Key Beach Mouse		X
Alabama Beach Mouse		X
Finback Whale		X
Humpback Whale		X
Blue Whale		X
Sei Whale		X
Sperm Whale		X
Hawksbill Sea Turtle		X
Leatherback Sea Turtle		X
Kemp's Ridley Sea Turtle		X
Green Sea Turtle		X
Loggerhead Sea Turtle	X	
Eastern Indigo Snake	X	

(This page intentionally left blank.)

Appendix C

Alabama Department of Environmental Management Coastal Sand Fencing Construction Guidelines

(This page intentionally left blank.)

ADEM Coastal Section Sand Fencing Construction Guidelines

The placement of sand fencing encourages the growth of sand dunes. These dunes in turn provide storm surge protection during storm and hurricanes and habitat for beach and dune plants and animals. However, sand fencing must be constructed in such a manner that impacts to nesting endangered sea turtles are minimized. The construction of sand fencing which is not designed to minimize impacts to nesting sea turtles may prevent female turtles from reaching nesting areas and may increase hatchling mortality by trapping hatchlings landward of the fencing. Therefore, in order to insure that sand fencing placed forward of the construction control line for dune enhancement purposes is constructed in such a manner that impacts to endangered sea turtles are minimized, the following guidelines and design criteria must be utilized:

1. Sand fencing must be constructed utilizing standard wood slat fencing commonly known as "sand fencing" or "snow fencing". Plastic fencing, silt fencing, and/or woven fabric fencing are not acceptable.

2. Supporting posts shall be no larger than 2" in width or 4" in diameter, placed by excavation, and shall not be secured by concrete.

3. Sand fencing shall be placed no farther seaward than the approximate seaward line of vegetation and/or in no case shall sand fencing be placed on the flat wet beach area seaward of the primary dune line.

4. Sand fencing shall be constructed in sections no longer than 10' in length spaced at a minimum of 7' apart on a diagonal alignment for the shore-parallel coverage of the subject property, as shown in the following diagram:

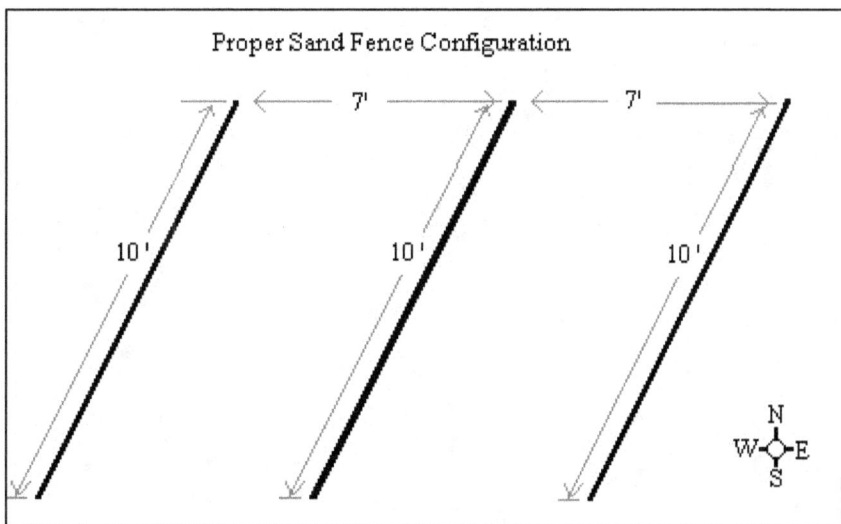

5. Persons wishing to obtain authorization to construct sand fencing seaward of the construction control line should submit to the Department the following information:
   A. the name, phone number and mailing address of the person wishing to construct the sand fencing;
   B. the street address, town and zip code of the site on which the sand fencing is to be constructed;
   C. the name of the person and/or contractor who will be installing the sand fencing;
   D. a drawing or site plan of the project showing the proposed configuration of the sand fencing and the sand fence's location relative to the construction control line, the seaward edge of vegetation and the water line; and
   E. a statement to the effect that the sand fencing will be constructed in accordance with this guidance.

   Approval of requests for authorization to construct of sand fencing can normally be provided by the Department within 1-2 working days of receipt. Prior to placing sand fencing or placing sand for dune enhancement purposes, the local building office must also be contacted to insure that the proper permits and/or approvals are obtained.

(This page intentionally left blank.)

Appendix D

Compliance with Other Potentially Applicable Laws and Regulations
(non-exclusive list)

(This page intentionally left blank.)

1. DOI regulations for implementing NEPA (43 C.F.R. Part 46)
2. Park System Resources Protection Act (16 U.S.C. § 19jj)
3. National Marine Sanctuaries Act (16 U.S.C. §§ 1431 et seq.)
4. Federal Water Pollution Control Act (33 U.S.C. §§ 1251 et seq.)
5. Endangered Species Act (16 U.S.C. §§ 1531 et seq.)
6. National Historic Preservation Act (16 U.S.C. 470 §§ et seq.)
7. Fish and Wildlife Conservation Coordination Act (16 U.S.C. §§ 661-666c)
8. Migratory Bird Treaty Act (16 U.S.C. §§ 703-712)
9. Migratory Bird Conservation Act (126 U.S.C. §§ 715 et seq.)
10. Coastal Zone Management Act (16 U.S.C. §§ 1451-1464)
11. Marine Mammal Protection Act (16 U.S.C. §§ 1361-1421h)
12. Magnuson-Stevens Fishery Conservation and Management Act (16 U.S.C. §§ 1801 et seq.)
13. Clean Air Act (42 U.S.C. §§ 7401 et seq.)
14. Rivers and Harbors Act (33 U.S.C. §§ 401, et seq.)
15. Safe Drinking Water Act (42 U.S.C. §§ 300f et seq.)
16. Noise Control Act (42 U.S.C. §§ 4901 et seq.)
17. Antiquities Act (16 U.S.C. §§ 431 et seq.)
18. Archaeological Resources Protection Act (16 U.S.C. §§ 470aa-470mm)
19. Native American Graves Protection and Repatriation Act (25 U.S.C. §§ 3001 et seq.)
20. Wild and Scenic Rivers Act (16 U.S.C. §§ 1271 et seq.)
21. Historic Sites Act (16 U.S.C. §§ 461-467)
22. Archaeological and Historic Preservation Act (16 U.S.C. §§ 469-469c)
23. Executive Order 11514, Protection and Enhancement of Environmental Quality (Mar. 5, 1970, as amended by Executive Order 11991 (May 24, 1977)
24. Executive Order 11593, Protection and Enhancement of the Cultural Environment (May 13, 1971)
25. Executive Order 11988, Floodplain Management (May 24, 1977)
26. Executive Order 11990, Protection of Wetlands (May 24, 1977)
27. Executive Order 12114, Environmental Effects Abroad of Major Federal Actions (Jan. 4, 1979)
28. Executive Order 12580 (Jan. 23, 1987), as amended by Executive Order 12777, Implementation of Section 311 of the Federal Water Pollution Control Act and the Oil Pollution Act (Oct. 19, 1991)
29. Executive Order 12898, Federal Actions to Address Environmental Justice in Minority Populations and Low-Income Populations (Feb. 11, 1994)
30. Executive Order 12962, Recreational Fisheries (June 7, 1995)
31. Executive Order 13007 – Indian Sacred Sites; and Executive Order 13175 – Consultation and Coordination with Indian Tribal Governments
32. Executive Order 13089, Coral Reef Protection (June 11, 1998)
33. Executive Order 13112, Invasive Species (Feb. 3, 1999)
34. Executive Order 13158, Marine Protected Areas (May 26, 2000)
35. Executive Order 13186, Responsibilities of Federal Agencies to Protect Migratory Birds (Jan. 17, 2001)

36. Executive Order 13352, Facilitation of Cooperative Conservation (Aug. 30, 2004)
37. Subpart G of the National Contingency Plan (40 C.F.R. §§ 300.600 et seq.)
38. White House Council on Environmental Quality regulations for implementing NEPA (40 C.F.R. §§1500 et seq.)
39. DOI Departmental Manual 516 and Environmental Statement Memoranda supplements
40. Anadromous Fish Conservation Act (AFCA) (16 USC §§ 757[a] et seq.)
41. Coastal Wetlands Planning, Protection and Restoration Act of 1990 (CWPPRA) (P.L. 101-646)
42. Energy Policy Act (Public Law 109-58, Section 384)
43. Water Resources Development Act (Public Law 110-114, Section 7001-7016)
44. Fish and Wildlife Conservation Act (16 USC §§ 2901 et seq.)
45. Information Quality Guidelines Issued Pursuant to Section 515 of P.L. 106-554
46. National Wildlife Refuge System Improvement Act of 1997 (NWRSIA) (16 USC § 668[dd])
47. Americans with Disabilities Act (P.L. 101-336)
48. Emergency Wetlands Resources Act (16 USC § 3901)
49. Estuarine Protection Act (16 USC §§ 1221 et seq.)
50. Marine Protection, Research, and Sanctuaries Act

Appendix E

NEPA Findings

(This page intentionally left blank.)

FINAL
ENVIRONMENTAL ASSESSMENT

LAKE HERMITAGE MARSH CREATION
BA-42

PLAQUEMINES PARISH, LOUISIANA

U.S. FISH AND WILDLIFE SERVICE

ECOLOGICAL SERVICES

LAFAYETTE, LOUISIANA

November 2011

FINAL
ENVIRONMENTAL ASSESSMENT

LAKE HERMITAGE MARSH CREATION
BA-42

PLAQUEMINES PARISH, LOUISIANA

November 2011

Preparer:
Kevin J. Roy
Senior Field Biologist

U.S. Fish and Wildlife Service
Ecological Services
646 Cajundome Blvd., Suite 400
Lafayette, Louisiana 70506

Phone: (337) 291-3100
Fax: (337) 291-3139

TABLE OF CONTENTS

FIGURES

TABLES

LAKE HERMITAGE MARSH CREATION
CWPPRA Project BA-42
Plaquemines Parish, Louisiana

SECTION 1.0 PURPOSE AND NEED FOR PROPOSED ACTION

SECTION 1.1 INTRODUCTION

Louisiana accounts for 90 percent of the coastal marsh loss in the lower 48 states (Dahl 2000). The most recent assessment of coastal land loss in Louisiana indicates an annual loss rate of approximately 15 square miles per year from 1985 to 2006 (Barras et al. 2008). Previous assessments indicated loss rates from approximately 25 square miles per year (Dunbar et al. 1992) to 35 square miles per year (Barras et al. 1994), and statewide coastal wetland loss is projected to be over 10 square miles per year from 2000 to 2050 (Barras et al. 2003). Causes of Louisiana's coastal wetlands loss include sea level rise, subsidence, sediment deprivation, canalization, saltwater intrusion, and altered hydrology (Turner and Cahoon 1987, Turner 1990). The wetland loss resulting from Hurricanes Katrina and Rita alone is estimated to be 198 square miles (Barras et al. 2008).

Concern over Louisiana's coastal wetland loss prompted President George Bush to sign into law the Coastal Wetlands Planning, Protection and Restoration Act (CWPPRA) in 1990. CWPPRA provides over $80 million per year for planning, design and construction of coastal restoration projects in Louisiana. Each year, a list of projects is selected for implementation and funds are approved for engineering and design. That annual list is referred to as the Priority Project List, and the Lake Hermitage Marsh Creation Project was funded as part of the 15th Priority Project List in 2006.

In 1998, the Louisiana Coastal Wetlands Conservation and Restoration Task Force (LCWCRTF) and the Wetlands Conservation and Restoration Authority (WCRA) developed the Coast 2050 Plan which serves as the official restoration plan for coastal Louisiana (LCWCRTF and WCRA 1998a). The Coast 2050 Plan divided the Louisiana coastal zone into four regions encompassing nine hydrologic basins, and restoration strategies were developed for each region. Each basin was also divided into mapping units for which additional strategies were developed. The Coast 2050 Plan would be implemented using a number of different funding sources including the CWPPRA, the Water Resources Development Act, and the State's Coastal Wetlands Conservation and Restoration Fund.

The Lake Hermitage Marsh Creation Project is located within Region 2, which encompasses the Barataria Basin, Breton Sound Basin, and Mississippi River Delta Basin. The project area is located in the eastern Barataria Basin, which is bounded by the Mississippi River on the east and Bayou Lafourche on the west (Figure 1). Wetlands in the upper part of the basin include swamp around Lake Des Allemands, fresh marsh around Lake Salvador, and isolated stands of bottomland hardwoods along relict distributary ridges such as Bayou Barataria. Intermediate marsh is encountered south of Lake Salvador, and extends southward to the northern shoreline of

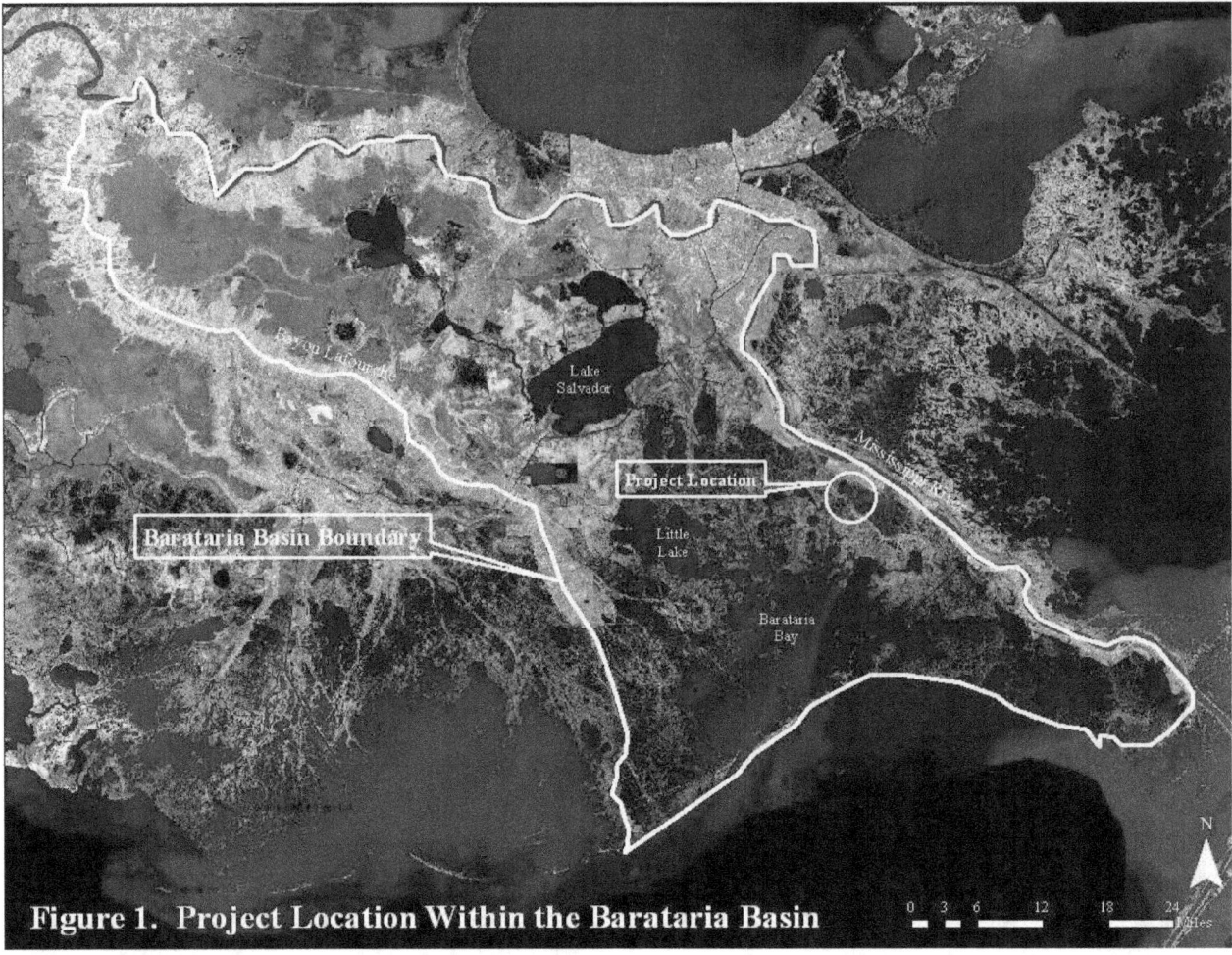

Figure 1. Project Location Within the Barataria Basin

Little Lake where brackish marsh becomes the dominant marsh type. Toward the northern edge of Barataria Bay, those marshes grade into saline marsh. A chain of barrier islands and barrier headlands separates the Barataria Basin from the Gulf of Mexico.

The project area is located along the eastern and southern shorelines of Lake Hermitage in the eastern Barataria Basin. The Jefferson Canal forms the southern boundary while the Bayou Grande Cheniere ridge forms the western boundary (Figure 2). Detailed drawings of all project features are found in Appendix A.

SECTION 1.2 PURPOSE OF PROPOSED ACTION

The purpose of the proposed project is to create emergent wetlands by hydraulically dredging sediments from the Mississippi River, and depositing that material in shallow open-water areas. In addition, Mississippi River sediments will be used to restore the eastern Lake Hermitage shoreline. The project area has experienced tremendous loss of emergent wetlands. Land-water data from the U.S. Geological Survey (USGS) indicates a 1985 to 2006 loss rate of -1.64 percent per year (U.S. Fish and Wildlife Service 2008). The causes of marsh loss appear to be primarily from subsidence and shoreline erosion from wind-generated waves. The need to address coastal

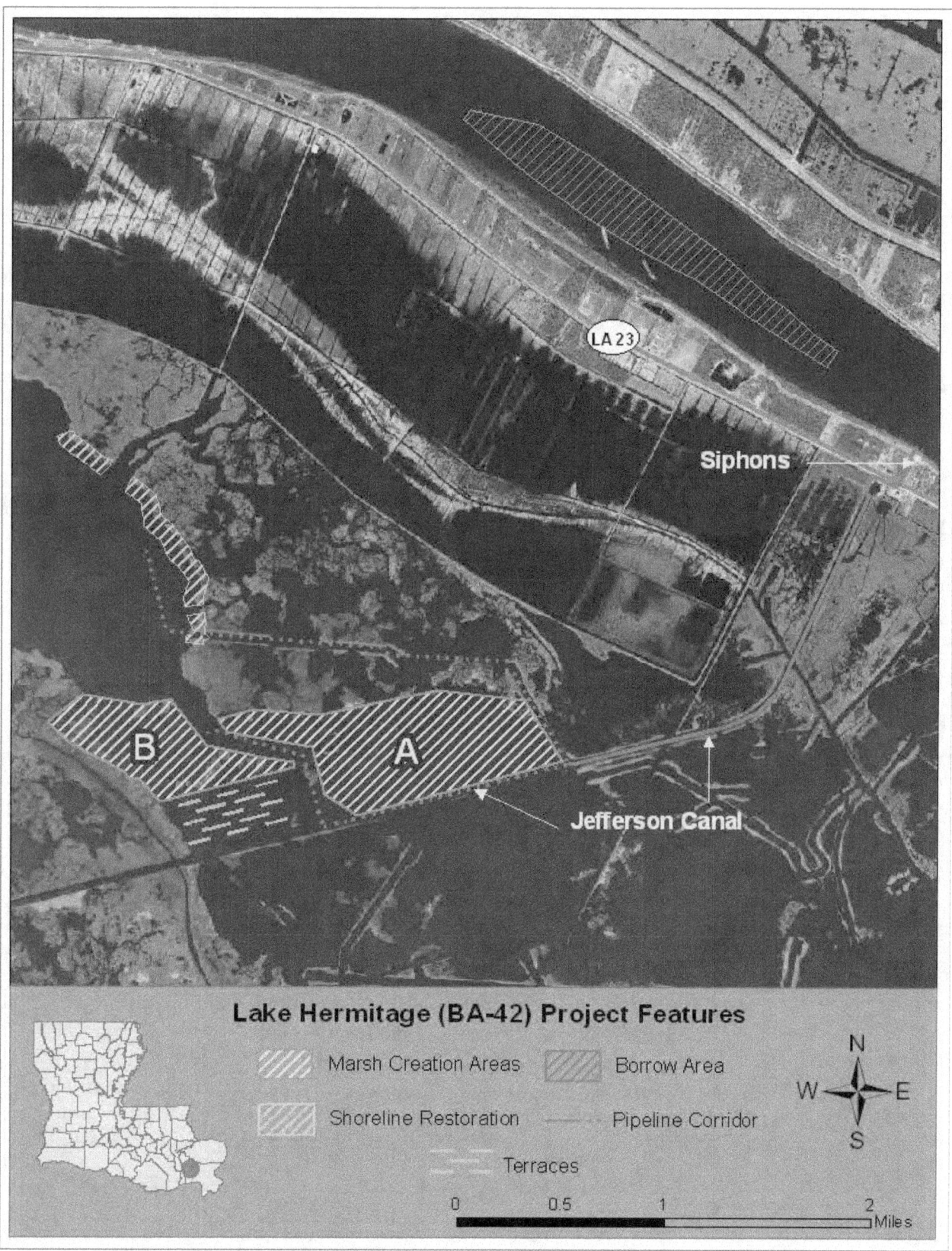

Figure 2. Lake Hermitage Marsh Creation Project Features (Lindquist 2008).

Louisiana's severe wetland loss has been identified in numerous restoration plans, programs, and State and Federal laws; implementation of the proposed project would help to fulfill that need.

The primary goals of the Lake Hermitage Marsh Creation Project are: 1) to restore the eastern Lake Hermitage shoreline to reduce erosion and prevent breaching into the interior marsh and 2) to re-create marsh in the open water areas south and southeast of Lake Hermitage. Specific goals of the project are to: 1) create 456 acres of marsh and nourish an additional 93 acres (Marsh Creation Areas A and B; Figure 2) by filling open-water areas and fragmented marsh with dredged material; 2) restore 7,400 linear feet (52 acres of marsh) of the eastern Lake Hermitage shoreline; and 3) create 6.5 acres of emergent habitat by constructing 7,300 linear feet of earthen terraces.

SECTION 1.3 PROBLEM

Historically, wetlands in the Barataria Basin were nourished by the fresh water, sediments, and nutrients delivered via overbank flooding of the Mississippi River and through its many distributary channels such as Bayou Lafourche, Bayou Barataria, and Bayou Grand Cheniere. As the flow of fresh water and sediments from the Mississippi River was restricted by flood protection levees and the closure of Bayou Lafourche, the basin began to gradually deteriorate from saltwater intrusion, subsidence, wave action, and sediment deprivation. From 1956 to 1990, the basin lost over 220,000 acres of marsh (Reed 1995) and from 1978 to 1990 it experienced the highest rate of wetland loss along the entire Louisiana coast (Barras et al. 2003).

The Coast 2050 Region 2 Plan divides the Barataria Basin into 21 mapping units or subbasins. The project area is located within the West Pointe a la Hache mapping unit (Figure 3), which contains approximately 19,000 acres of marsh and open water habitats (LCWCRTF and WCRA 1998*b*). Within the West Pointe a la Hache mapping unit, over 5,000 acres of wetlands were lost from 1932 to 1990. The primary causes of that loss were altered hydrology from canal dredging and subsidence. The rate of subsidence within this unit is high and ranges from 2.1 to 3.5 feet per century (LCWCRTF and WCRA 1998*b*).

The project area has experienced tremendous loss of emergent wetlands since 1956. Land-water data from the USGS indicates that over 1,100 acres of land were lost within the 1,600-acre project area from 1956 to 2006. The annual loss rate during that time period was over -2.6 percent per year. USGS land-water data for the West Pointe a la Hache mapping unit indicated a 1985-to-2006 loss rate of -1.64 percent per year (U.S. Fish and Wildlife Service 2008). The causes of marsh loss appear to be primarily from subsidence, altered hydrology, and shoreline erosion from wind-generated waves. Implementation of this project would create and protect important wetland habitat in the upper Lake Hermitage Basin. By offsetting the loss of emergent marsh and creating new marsh, fish and wildlife habitat quality and detrital production would increase.

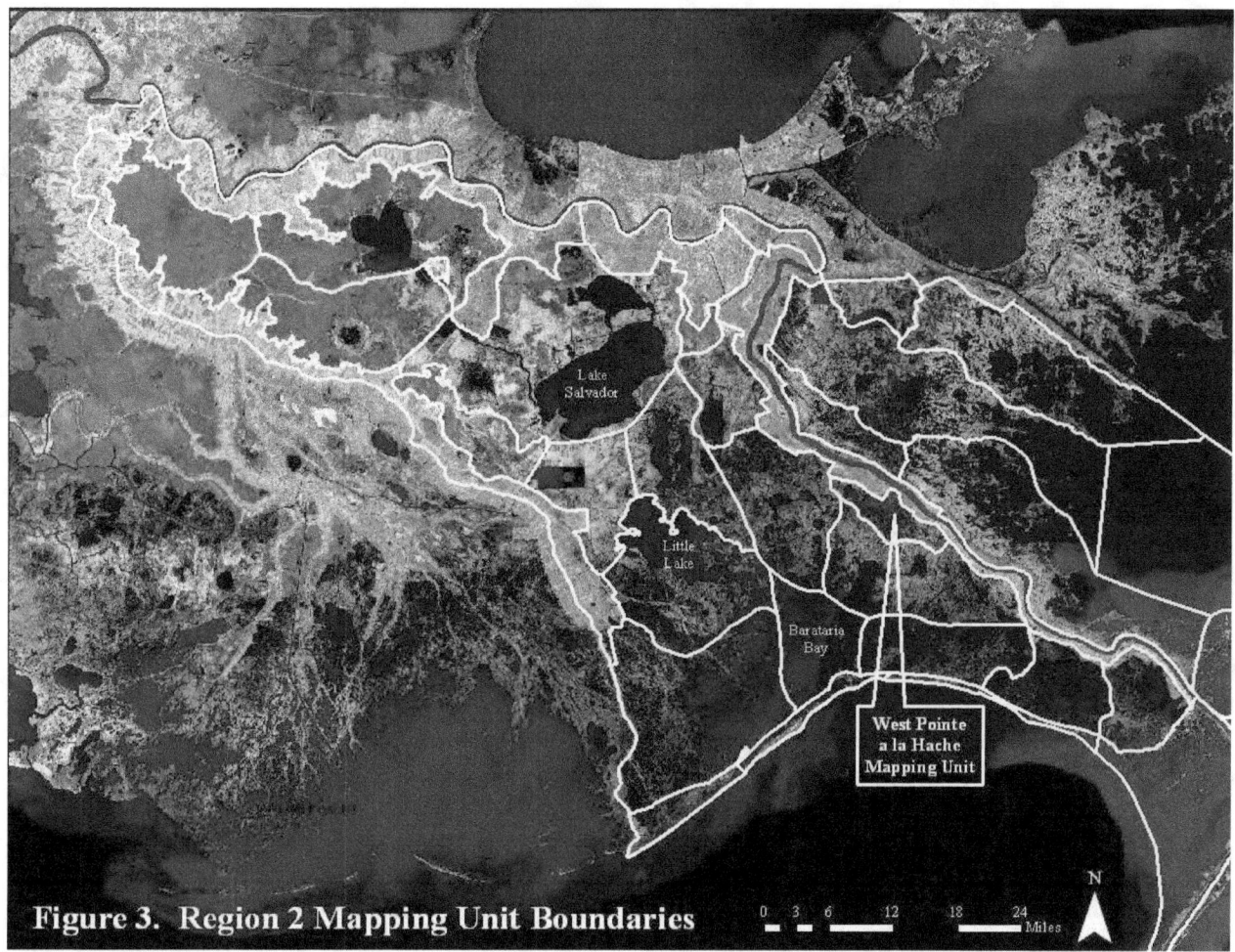

Figure 3. Region 2 Mapping Unit Boundaries

SECTION 1.4 REQUIRED DECISIONS

The decision to implement the Preferred Alternative has been made only after a thorough public review and full consideration of all comments. Opportunities for public comment occurred at public meetings conducted during the project development and selection stages of the CWPPRA planning process. Public meetings which offered the opportunity for public comment occurred on February 3, 2005, March 16, 2005, November 8, 2005, November 9, 2005, December 7, 2005, and February 8, 2006. Additional opportunities for public comment were provided during CWPPRA program meetings held on December 3, 2008 and January 21, 2009 when the project was approved for construction funding. Opportunity for public comment was also provided through review of the draft Environmental Assessment (EA) which was sent to the appropriate Federal, State, and local agencies, and other interested parties. Additional opportunity for public comment was provided during the application process for a Department of the Army Section 404 Clean Water Act permit. Upon review of all public and agency comments, the Service has determined that no further environmental documentation (e.g., Environmental Impact Statement) is necessary.

SECTION 1.5 COORDINATION AND CONSULTATION

Planning, engineering, and design of this project were coordinated with all LCWCRTF agencies, Plaquemines Parish, and other natural resource agencies. This project was nominated and selected as part of the 15th Priority Project List of CWPPRA. Projects on the 15th Priority Project List were nominated and developed at a series of public meetings held in February of 2005. Meeting participants included the LCWCRTF agencies, members of the CWPPRA Academic Advisory Group, landowners, environmental groups, Parish officials, and members of the general public. The CWPPRA Technical Committee met publicly on March 16, 2005, to consider preliminary costs and project benefits, and selected 6 projects for further evaluation as candidate projects. Interagency evaluations of those projects occurred from May to August 2005. Upon completion of project evaluations, public meetings were held on November 8 and 9, 2005, to allow the opportunity for public comment. The CWPPRA Technical Committee again met publicly on December 7, 2005, to select projects for recommendation to the CWPPRA Task Force. The CWPPRA Task Force selected 4 projects, including this one, for funding of engineering and design at a public meeting on February 8, 2006. Details concerning the plan formulation process for the 15th Priority Project List and the CWPPRA Standard Operating Procedures Manual are available at www.mvn.usace.army.mil/pd/cwppra_mission.utm.

An engineering and design review meeting was held on August 26, 2008, and a final design review meeting was held on November 3, 2008. All LCWCRTF agencies were invited to attend those meetings. The CWPPRA Technical Committee met publicly on December 3, 2008, when this project was selected for construction funding. The CWPPRA Task Force approved that selection at a public meeting held on January 21, 2009. Support for this project has been expressed by all entities involved.

SECTION 2.0   ALTERNATIVES INCLUDING THE PROPOSED ACTION

SECTION 2.1 ALTERNATIVE 1 - NO ACTION

Under this alternative, no action would be taken to restore the eastern Lake Hermitage shoreline or create marsh within the project area. Subsidence and interior marsh loss would continue to occur resulting in a decline in fish and wildlife productivity.

SECTION 2.2 ALTERNATIVE 2 – PREFERRED ALTERNATIVE

Project design information included within this section is taken from the Final (95%) Design Report prepared by the Louisiana Office of Coastal Protection and Restoration (LA OCPR 2008). Figure 2 displays the project features and detailed drawings of all project features are found in Appendix A.

The Preferred Alternative consists of hydraulically dredging bottom sediments in the Mississippi River and pumping that material into open-water and fragmented marsh areas in the project area to create approximately 549 acres of marsh. Containment dikes will be constructed around the fill sites to contain the dredged material slurry. Hydraulically-dredged river sediments will also be used to restore 7,400 linear feet of the Lake Hermitage shoreline resulting in the creation of approximately 52 acres of wetlands. In addition, 7,300 linear feet of earthen terraces will be constructed from *in situ* borrow material resulting in the creation of approximately 6.5 acres of wetlands. Approximately 246 acres of water bottom in the Mississippi River would be dredged to a maximum depth of -66 feet North American Vertical Datum of 1988 (NAVD 88; all following elevations are reported in NAVD 88).

Marsh Creation
Two marsh creation sites will be filled with hydraulically dredged material from the Mississippi River. To determine target elevations for the fill sites, marsh elevation surveys were performed. Marsh elevation surveys revealed that the average elevation of healthy marsh within the project area was approximately +1.2 feet (Sigma Consulting Group 2007). The mean water elevation for the project area is approximately 1.2 feet based on data from Coastwide Reference Monitoring System station 0260 for the period July, 2007 to June, 2011. Often, a goal of marsh restoration projects is for the marsh platform to settle to an elevation within the intertidal zone so that the created marsh functions similarly to natural marsh. To achieve a sustainable marsh elevation throughout the project life, the marsh platform will initially be pumped to a higher elevation during construction and allowed to settle to the desired target elevation over time.

Consolidation settlement calculations (Eustis Engineering Services, L.L.C. 2007) were also performed for borings taken within the fill sites to determine target elevations for the fill sites. The purpose of those calculations was to determine a fill elevation that would ultimately settle as close as possible to the existing healthy marsh elevation after 20 years. It was concluded that a target fill elevation of +2.0 feet would ultimately settle to an elevation of +1.3 feet. That value is extremely close to the existing healthy marsh elevation (+1.2 feet).

Containment dikes will be built to +3.0 feet with a 5-foot crown width and 1(V):6(H) side slopes. Containment dikes will be constructed with a bucket dredge using *in situ* material from within each fill site and the borrow area will be filled with hydraulically dredged material. It is anticipated that the containment dikes will subside and breach naturally to allow tidal connectivity and prevent ponding.

Shoreline Restoration
The shoreline restoration feature will consist of a sand fill template placed along the existing eastern Lake Hermitage shoreline. Hydraulically dredged material from the Mississippi River will be pumped along the shoreline to create this template. The shoreline restoration feature has been designed to maintain its integrity against the design wave height (+2.2 feet) based on the twenty year life of the project. Design parameters include a crown width of 50 feet, a lakeside slope of 1(V):50(H), and a marshside slope of 1(V):25(H). Design calculations indicated that the shoreline restoration feature should be constructed to an elevation of +4.2 feet to insure that a crown elevation of +2.2 feet NAVD is maintained throughout the twenty year life of the project. For constructability purposes, a crown elevation of +4.0 feet is proposed. Natural bayous along the shoreline will remain open. The shoreline slope will be planted with 4 rows (11,000 plugs)

of smooth cordgrass on 2.5-ft centers. The shoreline crown will be planted with 5 rows (7,400 four-inch containers) of seashore paspalum on 5-ft centers.

Terraces

A total of 7,300 linear feet of terraces will be constructed. Terraces will be approximately 500 to 700 feet long and built to an elevation of +3.5 feet with a 20-foot crown width and 1(V):3(H) side slopes. The terrace layout includes an overlapping configuration with a 250-foot spacing between terrace rows and 300 to 500-foot gaps between terraces. The terraces will be constructed with a bucket dredge using *in situ* material from within the terrace field. It is anticipated that several lifts will be required to obtain the desired elevation of +3.5 feet. The terrace slopes will be planted with three rows (17,000 plugs) of smooth cordgrass, on 2.5-ft centers. The perimeter of the terrace crowns will be planted with one row (4,000 four-inch containers) of seashore paspalum on 5-ft centers.

Alternative to Terraces

Additional project funding, from a non-CWPPRA source, may be available at the time of project construction to create an additional 104 acres of marsh instead of the above-mentioned 7,300 linear feet of terraces. The terrace field consists of approximately 104 acres so the alternative marsh creation would be constructed entirely within the existing the project boundary. Essentially, Marsh Creation Area B (Figure 2) would be expanded by 104 acres and encompass the entire terrace field. This alternative is noted as an "additive/deductive alternate" on Sheet 3 in Appendix A. The design for this alternative would follow the marsh creation design previously discussed.

Borrow Site

The proposed borrow site is located between Mississippi River Miles (RM) 49.5 and 52. This stretch of the river is located near the marsh fill site and depths are shallow enough to be reached using a large hydraulic dredge. Immediately upstream and downstream of this section, the water depths are too great to be dredged by a conventional dredge. The maximum depth of cut is assumed to be elevation -76.0 feet. The total volume of available sediment in this reach of the river is 9,421,546 cubic yards. The total fill volume required is 5,214,222 cubic yards, (including refilling containment dike borrow sites). Should the 104-acre marsh creation alternative be constructed, instead of the terraces, the total fill volume required would increase to 6,202,644 cubic yards.

The proposed borrow site also meets the following restrictions required by the U.S. Army Corps of Engineers (USACE):

• All excavations must be at least 750 feet from any protection levee centerline;
• Borrow sites must be outside the USACE maintained navigation channel;
• Excavation in the river must not be made less than 4,000 feet upstream of a bridge crossing;
• The side slopes of the borrow site must be no steeper than 1(V):5(H); and
• The excavation must proceed from landside to riverside limits to minimize the possibility of overburden failure of the bank.

Additionally, areas near or adjacent to concrete revetment mats were avoided. The western boundary of the borrow site is delineated by a 750 foot offset from the centerline of the Mississippi River levee. In this stretch of river, the navigation channel is located near the eastern bank, delineating the eastern boundary of the borrow site. Although the magnetometer surveys indicated the borrow site is free of known pipelines, the contractor will be required to perform a magnetometer survey prior to excavation.

Dredge pipeline crossing

The dredge slurry discharge pipeline will cross the Mississippi River levee near the West Pointe a la Hache Siphons on Plaquemines Parish property. A suitable levee crossing shall be built as per USACE's requirements. A casing will be installed underneath Highway 23 in accordance with all Louisiana Department of Transportation and Development specifications. From Highway 23, the pipeline will be placed on Plaquemines Parish property until it reaches the Jefferson Canal. It will then run parallel with the Jefferson Canal to the project area.

SECTION 2.3 OTHER ALTERNATIVES CONSIDERED

Other shoreline protection/restoration alternatives were considered for the eastern rim of Lake Hermitage. These alternatives consisted of an offshore rock dike and a rock dike placed on the shoreline. The shoreline protection feature proposed in Phase 0 included the placement of 6,000 feet of rip rap along the eastern shoreline of Lake Hermitage. Since Lake Hermitage has an average water depth of 4.6 feet, it is anticipated that approximately 2.6 miles of access channel would have to be dredged to mobilize rock barges to the project site. Additionally, the relatively mild wave climate in Lake Hermitage did not warrant the construction of a "hard" shoreline protection feature. The Project Team also investigated hydraulically pumping sand to restore the degraded shoreline. Using hydraulically pumped sand would not require the contractor to dig access and would not result in a significant increase in the mobilization cost of the project as the rock feature would have. The geotechnical analysis indicated that the sandy material in the borrow site would be suitable for constructing the shoreline restoration feature. For those reasons, the Project Team elected to move forward with the shoreline restoration feature using a sand fill template.

SECTION 3.0 AFFECTED ENVIRONMENT

SECTION 3.1 PHYSICAL ENVIRONMENT

A. Hydrology

The project area is located within an interdistributary basin between the Mississippi River and Bayou Grand Cheniere. Grand Bayou, another distributary of the Mississippi River, is also found within this basin and is an important tidal connection to the south. Tidal exchange also occurs through Bayou Hermitage located on the western side of Lake Hermitage and through oil and gas canals which dissect the Bayou Grand Cheniere ridge

The project area is predominantly a tidal, intermediate to brackish marsh with connectivity to Lake Hermitage via small tidal creeks and shoreline breaches. There is one oil and gas canal in the project area which provides a tidal connection between the interior marsh and Lake Hermitage. Freshwater inputs into the project area are provided by the West Pointe a la Hache siphons (maximum flow of 2,000 cfs), small forced drainage pump stations, and rainfall.

B. Water Quality

The Louisiana Department of Environmental Quality (LDEQ) surface water monitoring program is designed to measure progress towards achieving water quality goals at the state and national levels, to gather baseline data used in establishing and reviewing the state water quality standards, and to provide a database for use in determining the assimilative capacity of the waters of the State. The surface water monitoring program consists of a fixed station long-term network, intensive surveys, special studies, and wastewater discharge compliance sampling. The LDEQ routinely monitors 29 conventional parameters and fecal coliform bacteria on a monthly or bimonthly basis using a fixed station, long-term network. In addition to the conventional parameters, volatile organic compounds are sampled at each site (Louisiana Department of Environmental Quality 2004).

The Louisiana Water Quality Standards define eight designated uses for surface waters: primary contact recreation, secondary contact recreation, fish and wildlife propagation, drinking water supply, shellfish propagation, agriculture, outstanding natural resource, and limited aquatic and wildlife use. Each water body is evaluated as fully supporting, partially supporting, or not supporting of each of its designated use(s). No water quality assessments are available for any water body within the project area. Water quality assessments for the Wilkinson Canal and Wilkinson Bayou, approximately five miles west of the project area, are presented in Table 1. Both waterbodies are listed as fully supporting their designated uses.

Table 1. Evaluation of water quality (LDEQ 2006).

Water Body Subsegment Code	Water Body Name and Description	Primary Contact Recreation	Secondary Contact Recreation	Fish and Wildlife Propagation
LA020904	Wilkinson Canal and Wilkinson Bayou	Fully Supporting	Fully Supporting	Fully Supporting

SECTION 3.2 BIOLOGICAL ENVIRONMENT

A. Vegetation

The project area was classified as brackish three-cornered grass marsh in 1949 (O'Neil 1949). Coastal vegetative type maps in 1968 (Chabreck et. al., 1968), 1978 (Chabreck and Linscombe 1978), and in 1988 (Chabreck and Linscombe 1988) classified the area as brackish marsh. The 1997 survey (Chabreck and Linscombe 1997) classified the entire project area as intermediate and the 2001 survey (Linscombe and Chabreck 2001) classified the area as approximately 50% brackish marsh and 50% intermediate marsh. However, the 2007 survey (Sasser et. al., 2007) classified the entire project area as saline marsh. Plant communities observed during field investigations indicate that the project area supports brackish marsh dominated by marshhay cordgrass. Other common species include eastern baccharis, Olney bulrush, smooth cordgrass, and deerpea. The northern extremes of the project area contain more of an intermediate marsh community and more diverse submerged aquatic vegetation.

Based on recent habitat classification data and field observations, the project area appears to lie in a transition zone between intermediate and brackish marsh. Intermediate to low-salinity brackish conditions likely prevail during high rainfall years and prolonged operation of the West Pointe a la Hache siphons. Brackish and sometimes saline conditions are likely to prevail during low rainfall years and periods of inconsistent siphon operation. During siphon operation, salinities often remain below 2 parts per thousand (ppt) and the average annual salinity from 1992-2002 for two monitoring stations within the project area was 5 ppt (Boshart 2003).

B. Fisheries

The project area supports a diverse assemblage of estuarine-dependent fishes and shellfishes, and species presence is largely dictated by salinity levels and season. During low-salinity periods, species such as Gulf menhaden, blue crab, white shrimp, and striped mullet are present in the project area. During high-salinity periods, more salt-tolerant species such as spotted seatrout, black drum, red drum, Atlantic croaker, sheepshead, southern flounder, and brown shrimp may move into the project area. Wetlands throughout the project area also support small resident fishes and shellfish such as least killifish, sheepshead minnow, sailfin molly, grass shrimp and others. Those species are typically found along marsh edges or among submerged aquatic vegetation, and provide forage for a variety of fish and wildlife.

The proposed borrow site lies within the Mississippi River which provides habitat for an incredible diversity of freshwater fisheries many of which are commercially and recreationally important. Common species include gizzard shad, common carp, channel catfish, blue catfish, freshwater drum, smallmouth buffalo, white bass, and river shiner.

C. Essential Fish Habitat

The project is located within an area identified as Essential Fish Habitat (EFH) by the Magnuson-Stevens Fishery Conservation and Management Act (MSFCMA). The 1998 generic amendment of the Fishery Management Plans for the Gulf of Mexico, prepared by the Gulf of Mexico Fishery Management Council, identifies EFH in the project area to be estuarine emergent wetlands, submerged aquatic vegetation (SAV), estuarine water column, and mud substrates. Under the MSFCMA, wetlands and associated estuarine waters in the project area are identified as EFH for postlarval/juvenile and subadult brown shrimp; postlarval/juvenile and subadult white shrimp; and postlarval/juvenile, subadult, and adult red drum. Table 2 provides a more detailed description of EFH within the project area.

Table 2. EFH Requirements for Managed Species that Occur in the Project Area.

Species	Life Stage	Essential Fish Habitat	Occurrence in Project Area
Brown shrimp	postlarval/juvenile	marsh edge, SAV, tidal creeks, inner marsh	All habitats are found throughout the project area
	subadult	mud bottoms, marsh edge	All habitats are found throughout the project area
White shrimp	postlarval/juvenile subadult	marsh edge, SAV, marsh ponds, inner marsh, oyster reefs	All habitats are found throughout the project area (excluding oyster reefs)

Red drum	postlarval/juvenile	SAV, estuarine mud bottoms, marsh/water interface	All habitats are found throughout the project area
	subadult	mud bottoms, oyster reefs	Mud bottoms are found within open-water areas
	adult	Gulf of Mexico & estuarine mud bottoms, oyster reefs	Estuarine mud bottoms are found within open-water areas

D. Wildlife

The project area provides important habitat for several species of wildlife, including waterfowl, wading birds, shorebirds, mammals, reptiles and amphibians. The project area provides wintering habitat for migratory puddle ducks including gadwall, blue-winged teal, green-winged teal, American widgeon, and northern shoveler. Diving duck species which utilize the project area include lesser scaup and ring-necked ducks. The resident mottled duck, which nests in fresh to brackish marshes, is found throughout the year.

Common wading bird species which utilize the project area include the great blue heron, green heron, tricolored heron, great egret, snowy egret, yellow-crowned night-heron, black-crowned night-heron, and white ibis. Mudflats and shallow-water areas provide habitat for numerous species of shorebirds and seabirds. Shorebirds include the American avocet, willet, black-necked stilt, dowitchers, and various species of sandpipers. Seabirds include the white pelican, herring gull, laughing gull, and several species of terns.

Migratory and resident non-game birds, such as the boat-tailed grackle, red-winged blackbird, seaside sparrow, northern harrier, belted kingfisher, and marsh wrens, also utilize the project area. Important gamebirds found in the area include the clapper rail, sora rail, Virginia rail, American coot, common moorhen, and common snipe in addition to resident and migratory waterfowl.

Mammals found within the project area include nutria, muskrat, mink, river otter, and raccoon, all of which are commercially important furbearers. Reptiles and amphibians are fairly common in the low-salinity brackish and intermediate marshes found within the project area. Reptiles include the American alligator, western cottonmouth, water snakes, speckled kingsnake, rat snake, and eastern mud turtle. Amphibians expected to occur in the area include the bullfrog, southern leopard frog, and Gulf coast toad.

E. Threatened and Endangered Species

The pallid sturgeon is an endangered fish found in Louisiana, in both the Mississippi and Atchafalaya Rivers (with known concentrations in the vicinity of the Old River Control Structure Complex); it is possibly found in the Red River as well. The pallid sturgeon may be found within the proposed borrow site for this project which is located within the Mississippi River. The pallid sturgeon is adapted to large, free-flowing, turbid rivers with a diverse assemblage of physical characteristics that are in a constant state of change. Detailed habitat requirements of this fish are not known, but it is believed to spawn in Louisiana. Habitat loss through river channelization and dams has adversely affected this species throughout its range. Entrainment issues associated with dredging operations in the Mississippi and Atchafalaya Rivers and through diversion structures off the Mississippi River are two potential effects that should be addressed in future planning studies and/or in analyzing current project effects.

Federally listed as an endangered species, West Indian manatees occasionally enter Lakes Pontchartrain and Maurepas, and associated coastal waters and streams during the summer months (i.e., June through September). Manatee occurrences appear to be increasing, and they have been regularly reported in the Amite, Blind, Tchefuncte, and Tickfaw Rivers, and in canals within the adjacent coastal marshes of Louisiana. They have also been occasionally observed elsewhere along the Louisiana Gulf coast. The manatee has declined in numbers due to collisions with boats and barges, entrapment in flood control structures, poaching, habitat loss, and pollution. Cold weather and outbreaks of red tide may also adversely affect these animals.

SECTION 3.3 CULTURAL AND RECREATIONAL RESOURCES

Various cultural resources occur throughout the Louisiana coastal zone, including both prehistoric and historic sites. The Louisiana Department of Culture, Recreation and Tourism maintains catalogues of cultural resource sites, but many areas remain unsurveyed and the significance or eligibility of some sites for inclusion in the National Register of Historic Places has not been determined. A review by the Louisiana Office of Cultural Development, Division of Archeology indicated that no archaeological sites are located within the project area. The Louisiana Office of Cultural Development has indicated, by correspondence dated October 6, 2008, that they have no objections to project implementation.

Recreational use of the project area is oriented primarily toward hunting, fishing, and non-consumptive uses such as wildlife observation. Access to the project area is by boat only, as no roads or highways are present.

SECTION 3.4 ECONOMIC RESOURCES

Project-area wetlands provide essential nursery habitat for commercially and recreationally important fishes and shellfishes such as Gulf menhaden, red drum, spotted seatrout, southern flounder, brown shrimp, white shrimp, blue crab and others. National Marine Fisheries Service statistics for the last 20 years indicate that coastal Louisiana contributes approximately 20 percent of the nation's total commercial fisheries harvest (LCWCRTF and WCRA 1998a). In 2003, commercial fishery landings in coastal Louisiana exceeded 1 billion pounds with a dockside value of over $285 million with a total economic effect of more than $2.5 billion (Southwick Associates 2005). Additionally, Louisiana's shrimp and oyster harvests comprise approximately 35 to 40 percent of the national total for those species (LCWCRTF 1993).

Louisiana's coastal wetlands also produce more wild furs and alligator skins than any other State in the nation. Nutria, muskrat, and raccoon constitute 94 percent of the value of the Louisiana fur industry, valued at approximately $1.3 million annually (Louisiana Fur and Alligator Advisory Council 1997). In 2003, the Louisiana fur harvest totaled $1.6 million (Southwick Associates 2005). The wild alligator harvest is also an important economic resource in coastal Louisiana. The wild harvest from 1972 to 1997 produced one million skins with an estimated value of $128.6 million. The annual harvest averaged 26,742 from 1992 to 1997, and the value of skins and meat was worth over $9.3 million (Louisiana Fur and Alligator Advisory Council

1997) during that period. In 2003, the wild alligator harvest totaled over $6 million in retail sales (Southwick Associates 2005).

Recreational saltwater fishing contributed over $435 million to Louisiana's economy in 2003 (Southwick Associates 2005). Coastal marshes also provide substantial economic value associated with waterfowl hunting.

SECTION 4.0   ENVIRONMENTAL CONSEQUENCES

SECTION 4.1 ALTERNATIVE 1 - NO ACTION

A. Physical Environment

Hydrology
Under the No Action Alternative, the hydrology of the project area would likely be altered by the ongoing processes of shoreline erosion, shoreline breaching, and marsh deterioration. As marsh loss continues, tidal connectivity with Lake Hermitage and large expanses of open water south of the project area could increase as more tidal channels form and tidal exchange increases. In several sections of the project area along the Lake Hermitage shoreline, the shoreline rim is very narrow between the lake and interior marsh ponds. Continued shoreline erosion will likely result in more breaches forming between the lake and interior ponds.

Water Quality
Under the No Action Alternative, water quality in the project area will likely remain the same.

B. Biological Environment

Vegetation
Under the No Action Alternative, vegetation in the project area would likely remain the same as it is today with vegetation typical of a brackish marsh. Marshhay cordgrass would likely remain as the dominant plant species.

Marsh loss from shoreline erosion and subsidence would continue. The Wetland Value Assessment (WVA) prepared by the CWPPRA Environmental Work Group projected that shoreline erosion would continue at approximately 12.7 feet per year and interior marsh loss would continue at a rate of -1.64 percent per year, resulting in the loss of 126 acres of marsh (USFWS 2008).

Fisheries
Although marsh loss would continue under the No Action Alternative, the project area would continue to support a diverse assemblage of estuarine-dependent fishery species. However, the loss of intertidal, emergent wetlands to shallow, unvegetated open water would result in decreased fishery productivity. As a marsh complex exceeds 70 percent unvegetated open water, shrimp and blue crab populations may decline (Minello and Rozas 2002).

The proposed borrow site within the Mississippi River would continue to support a diverse assemblage of freshwater fish species.

Essential Fish Habitat Assessment

Under the No Action Alternative, estuarine marsh is the primary type of EFH impacted by continued wetland loss and deterioration. According to the WVA conducted by the CWPPRA Environmental Work Group, 126 acres of emergent marsh would be converted to shallow open water (i.e., mud bottom) over the project life. Although an increase in some types of EFH (i.e., mud bottom and estuarine water column) would occur, adverse impacts would occur to more productive types of EFH (i.e., estuarine emergent wetlands). The loss of estuarine emergent wetlands would result in negative impacts to postlarval/juvenile and subadult brown shrimp; postlarval/juvenile and subadult white shrimp; and postlarval/juvenile red drum.

Coverage of submerged aquatic vegetation, another important type of EFH, is projected to decrease slightly over the project life from 9 percent coverage of the open water areas to 5 percent coverage (USFWS 2008) as marsh loss continues.

Wildlife

Under the No Action Alternative, the project area would continue to provide habitat for a multitude of species including migratory waterfowl, wading birds, shorebirds, mammals, reptiles, and amphibians. However, the continued loss of emergent wetlands would negatively impact those species which utilize the project area. The intertidal marsh and shallow, isolated ponds and associated submerged aquatic vegetation are utilized by those species for foraging, resting, or nesting habitat. Conversion of that habitat type to unvegetated, open-water areas would diminish habitat value for all wildlife species.

Threatened and Endangered Species

The endangered pallid sturgeon is found in both the Mississippi and Atchafalaya Rivers and may be found within the proposed borrow site for this project. Use of that area by the pallid sturgeon would likely continue under the No Action Alternative.

The endangered West Indian manatee is occasionally found in Lakes Pontchartrain and Maurepas, and associated coastal waters and streams during the summer months (i.e., June through September). Manatee occurrences appear to be increasing, and they have been regularly reported in the Amite, Blind, Tchefuncte, and Tickfaw Rivers, and in canals within the adjacent coastal marshes of Louisiana. They have also been occasionally observed elsewhere along the Louisiana Gulf coast. Although unlikely to occur in the project area, their use would continue under the No Action Alternative.

C. Cultural and Recreational Resources

No archeological sites are located within the project area; therefore, no impacts are expected under the No Action Alternative. Recreational opportunities within the project area, such as hunting and fishing, may decrease somewhat with the ongoing loss of marsh and diminished capacity of the area to support fish and wildlife populations.

D. Economic Resources
Commercial and recreational activities within the project area are important components of the local economy. Waterfowl hunting, recreational fishing, and commercial shrimping and crabbing contribute greatly toward the economies of the surrounding communities. The continued loss of emergent wetlands would decrease the project area's ability to support those activities.

SECTION 4.2 ALTERNATIVE 2 - PREFERRED ALTERNATIVE

A. Physical Environment

Hydrology
Under the Preferred Alternative, hydrologic conditions within the project area would be impacted by the creation of marsh and restoration of the eastern Lake Hermitage shoreline. The large, open-water areas and some of the tidal waterways through which water exchange now occurs would be filled with dredged material. However, tidal connectivity between the project area and Lake Hermitage would be maintained. Two natural tidal channels which connect the interior marsh to Lake Hermitage would not be filled. In addition, tidal channels are anticipated to form as differential settlement of dredged material occurs throughout the marsh creation areas. Existing tidal channels, boat trails, and other waterways occur throughout the project area, and higher settlement of dredged material is anticipated in those areas, because they are deeper than the adjacent open-water areas to be filled. Those areas would be the lowest points on the marsh platform, so water exchange would naturally occur at those sites. In addition, the marsh platform is anticipated to consolidate and settle to the existing marsh elevation over the project life. As the marsh platform subsides, more tidal connections and other open-water areas would form throughout the project area.

Water Quality
Under the Preferred Alternative, dredging activities in the Mississippi River, the placement of dredged material in the project area, and the construction of containment dikes and terraces would increase turbidity as bottom sediments are disturbed. However, the increased turbidity would only occur during periods of active dredging and is expected to dissipate rapidly upon completion of construction. Dewatering of the marsh creation fill sites will also result in increased turbidities in the surrounding open water areas. In addition, turbidities may increase after rainfall events as water runs off the unvegetated marsh platform, especially immediately after dredged material deposition.

B. Biological Environment

Vegetation
Under the Preferred Alternative, approximately 549 acres of marsh would be created/nourished within the marsh creation cells and an additional 52 acres of marsh will be created as part of the shoreline restoration feature. In addition, 6.5 acres of emergent habitat would result from construction of the earthen terraces. Very little emergent vegetation would be present immediately after construction as most of the project area would be unvegetated dredged

material. Those areas of marsh which are nourished would likely revegetate more rapidly than the large, open-water areas which are filled. Marsh vegetation nourished with 6 to 12 inches of material has been shown to respond favorably and revegetate quickly (Mendelssohn and Kuhn 1999). Large, open-water areas which are filled with dredged material would likely revegetate at a slower rate than nourished marsh. However, based on the performance of other marsh creation projects, revegetation could be expected within 1 to 2 years after construction. Operation of the West Pointe a la Hache siphons, which will provide fresh water and nutrients to the project area, would enhance conditions for vegetative colonization. Vegetative communities would likely be very similar to those currently found within the project area and marshhay cordgrass would likely remain as the dominant species.

Under the Preferred Alternative, marsh loss would continue in the project area, but at a reduced rate. The WVA prepared by the CWPPRA Environmental Work Group projected that land loss would continue at a rate of -0.82 percent per year, compared to -1.64 percent per year under the No Action Alternative (USFWS 2008). Within the project area, 702 acres of marsh would remain at the end of the 20-year project life compared to 255 acres under the No Action Alternative, and a substantial acreage of marsh would remain within the project area for many years beyond the project life.

Should the 104-acre marsh creation alternative be constructed, instead of the terraces, then 653 acres of marsh would be created/nourished along with the 52 acres resulting from the shoreline restoration. A total of 785 acres of marsh would remain at the end of the 20-year project life compared to 255 acres under the No Action Alternative.

The WVA indicates that the coverage of submerged aquatic vegetation is also projected to increase from 9 percent of the open-water areas to 25 percent (USFWS 2008). The smaller, shallower ponds which would form within the marsh platform would be more conducive for the establishment of submerged aquatic vegetation. Those smaller waterbodies would be less susceptible to increases in turbidity from wind-generated waves. In addition, reduced tidal connectivity would enhance the growth of submerged aquatics.

Fisheries
Under the Preferred Alternative, the project area would continue to support a diverse assemblage of fishes and shellfishes. The creation and nourishment of intertidal marsh would ensure that the project area continues to provide important nursery functions well beyond the 20-year project life. Several studies indicate that vegetated habitats (i.e., emergent marsh and submerged aquatic vegetation beds) generally support higher densities of fish and crustaceans than unvegetated habitat (Castellanos and Rozas 2001, Rozas and Minello 2001, Minello and Rozas 2002). Population declines of shrimp and blue crabs may become evident when a marsh complex exceeds 70 percent unvegetated, open water (Minello and Rozas 2002). Compared to the No Action Alternative, an additional 447 acres of marsh would result from project implementation (USFWS 2008). Much of that habitat would exist within the intertidal zone and would provide foraging and nursery habitat for a number of estuarine species.

The marsh creation alternative for the terrace field would increase the amount of intertidal marsh within the project area. An additional 104 acres of marsh would be created instead of the 6.5

acres of emergent habitat which would result from the terraces. Although the terraces would provide a significant amount of edge habitat for fish and shellfish species, the longevity of the 104 acres of created marsh would significantly exceed that of the terraces. Compared to the No Action Alternative, an additional 530 acres of marsh would result over the project life.

Dredging activities in the Mississippi River would increase turbidity as bottom sediments are disturbed. The increased turbidity and disturbance from dredging activities could result in some fishery species being displaced. It is likely that those species would simply relocate to an area of more suitable habitat. However, the increased turbidity would only occur during periods of active dredging and is expected to dissipate rapidly once dredging activities cease.

Essential Fish Habitat Assessment
Estuarine emergent wetland is the primary type of EFH that would increase significantly under the Preferred Alternative; such habitat would be created in open-water areas and deteriorated marsh. According to the WVA, 447 additional acres of emergent marsh would exist at the end of the project life under the Preferred Alternative, compared to the No Action Alternative. With the alternative construction of 104 acres of marsh within the terrace field, the net gain in marsh acreage increases from 447 to 530. Coverage of submerged aquatic vegetation is also expected to increase. Increases in those habitat types would benefit postlarval/juvenile and subadult brown shrimp; postlarval/juvenile and subadult white shrimp; and postlarval/juvenile red drum.

The creation of estuarine emergent wetlands would result in the loss of mud bottom and estuarine water column as emergent marsh would replace those habitat types. Loss of mud bottom EFH could result in negative impacts to subadult brown shrimp and postlarval/juvenile, red drum. Although adverse impacts would occur to some types of EFH, more productive types of EFH (i.e., estuarine emergent wetlands) would be created under the Preferred Alternative. In addition, open-water habitat would form within the marsh platform as ponds and other waterbodies develop as a result of natural marsh loss processes. Open-water habitats are expected to contain 25 percent coverage of submerged aquatic vegetation compared to only 5 percent coverage under the No Action Alternative. Therefore, the Preferred Alternative would result in a net positive benefit to all managed species that occur in the project area.

Wildlife
The Preferred Alternative would result in improved habitat conditions for several species of wildlife including migratory and resident waterfowl, shorebirds, wading birds, and furbearers. Migratory waterfowl utilizing the project area would benefit from a greater food supply resulting from the increased abundance and diversity of emergent and submerged species. Habitat for the resident mottled duck would also improve considerably as the marsh platform, shoreline berm, and terraces would provide more desirable nesting habitat.

Intertidal marsh and marsh edge would also provide increased foraging opportunities for shorebirds and wading birds. Small fishes and crustaceans are often found in greater densities along vegetated marsh edge (Castellanos and Rozas 2001, Rozas and Minello 2001), and many of those species are important prey items for wading birds such as the great blue heron, little blue heron, great egret, black-crowned night-heron, and snowy egret. Mudflats and shallow water habitat created by the deposition of dredged material would provide increased foraging opportunities for shorebirds such as least sandpipers, killdeer, and the American avocet. Those species feed on tiny invertebrates and crustaceans found on mudflats which are exposed at low tide and in shallow-water areas of the appropriate depth.

Furbearers (such as the nutria and muskrat) which feed on vegetation would benefit from the increased marsh acreage in the project area. Representative furbearers such as the mink, river otter, and raccoon have a diverse diet and feed on many different species of fishes and crustaceans. Those species often feed along vegetated shorelines which provide cover for many of their prey species.

Threatened and Endangered Species
The Service has conducted an Intra-Service Section 7 Endangered Species Act consultation of the Preferred Alternative's impacts on the pallid sturgeon and West Indian Manatee. Based on that consultation, the Service has determined that the Preferred Alternative would be "not likely to adversely affect" the endangered pallid sturgeon and endangered West Indian manatee.

The pallid sturgeon is known to inhabit the waters of the Mississippi River and may be found within the designated borrow area. To ensure protection of the pallid sturgeon, all personnel associated with the project will be informed of the potential presence of the pallid sturgeon and take actions to induce them to leave the immediate work area prior to dredging regardless of water depth or time of year. Specific language has been included within the project's plans and specifications to avoid/minimize impacts to the pallid sturgeon. The following actions shall be implemented to help prevent any potential project related direct or indirect effects to the pallid sturgeon:

> 1) The hydraulic dredge cutterhead shall remain completely buried in the bottom material during dredging operations.
>
> 2) If pumping water through the cutterhead is necessary to dislodge material or to clean the pumps or cutterhead, etc., the pumping rate shall be reduced to the lowest rate possible until the cutterhead is at mid-depth, where the pumping rate can then be increased.
>
> 3) During dredging, the pumping rates shall be reduced to the slowest speed feasible while the cutterhead is descending to the channel bottom.

The West Indian manatee, although it is unlikely, may be found in the estuarine waters in or near the project area. Construction equipment (e.g., boats, barges, airboats) may encounter manatees in the waterbodies found within and around the project area. Specific language has been included within the project's plans and specifications to avoid/minimize impacts to the West Indian manatee. The following precautions will be implemented from May to October, when manatees have the greatest potential for entering the project area:

> 1) All construction personnel are responsible for observing water-related activities for the presence of manatee(s) which are protected under the Marine Mammal Protection Act of 1972 and the Endangered Species Act of 1973.

2) All personnel associated with the project shall be instructed about the possible presence of manatees and the need to avoid collisions with and injury to manatees. Any sighting of, collision with, or injury to a manatee shall be immediately reported to the Engineer.

Temporary signs should be posted prior to and during all construction/dredging activities to remind personnel to be observant for manatees during active construction/dredging operations or within vessel movement zones (i.e., work area), and at least one sign should be placed where it is visible to the vessel operator. Siltation barriers, if used, should be made of material in which manatees could not become entangled, and should be properly secured and monitored. If a manatee is sighted within 100 yards of the active work zone, special operating conditions should be implemented, including: no operation of moving equipment within 50 feet of a manatee; all vessels should operate at no wake/idle speeds within 100 yards of the work area; and siltation barriers, if used, should be re-secured and monitored. Once the manatee has left the 100-yard buffer zone around the work area on its own accord, special operating conditions are no longer necessary, but careful observations would be resumed. Any manatee sighting should be immediately reported to the Service's Lafayette, Louisiana Field Office (337/291-3100) and the Louisiana Department of Wildlife and Fisheries, Natural Heritage Program (225/765-2821).

C. Cultural and Recreational Resources
By correspondence dated October 6, 2008, the Louisiana Department of Culture, Recreation and Tourism indicated that no archaeological sites are located within the project area and, therefore, they have no objection to implementation of the Preferred Alternative.

Recreational opportunities within the project area, such as hunting, fishing, and bird watching, may increase with the increased formation of emergent marsh and other fish and wildlife habitats. An increase in habitat value would likely result in increased fish and wildlife usage of the project area.

D. Economic Resources
By increasing emergent wetlands, and subsequently fish and wildlife resources, the Preferred Alternative would help to maintain that portion of the local economy dependent on recreational and commercial fish and wildlife resources found within the project area. Project-area waterfowl hunting and recreational fishing are important components of the local economy, and creation of emergent marsh and other fish and wildlife habitats could increase the ability of the project area to support those activities. The increased acreage of emergent wetlands would also act as a storm buffer for flood protection levees north and east of the project area.

SECTION 5.0 RATIONALE FOR SELECTING PREFERRED ALTERNATIVE

Marsh loss in the project area has resulted in a decline in fish and wildlife habitat. Marsh loss is likely to continue in the project area at current rates and may increase as more breaches occur along the Lake Hermitage shoreline. Marsh elevations in some areas of deteriorated marsh are not conducive to the continued existence of the dominant plant species, marshhay cordgrass, which prefers higher elevations. Ponding and prolonged inundation, due to subsidence, have resulted in the deterioration of marsh and the formation of shallow, open-water habitat. Continued subsidence would result in the future deterioration of the remaining stands of healthy, unfragmented marsh. Elevation surveys conducted at three sites within the project area indicate an average marsh elevation of +1.2 feet (Sigma Consulting Group 2007). With the current design elevation of +2.0 feet, the marsh platform would support emergent vegetation throughout the 20-year project life.

Dedicated dredging to create marsh in shallow, open-water areas has been successfully used as a restoration technique across coastal Louisiana. Since CWPPRA was authorized in 1990, several marsh creation projects have been constructed and many more are authorized for engineering and design, or construction, by the LCWCRTF (Table 3) (Lindquist and Martin 2007). Also, several barrier island restoration projects have been constructed which utilize hydraulic dredging to create dune and marsh habitats. In addition, many other marsh creation projects have been constructed by the State of Louisiana through its Coastal Restoration Program as mitigation for wetland impacts under Section 404 of the Clean Water Act, and by the Corps of Engineers under other authorities such as Sections 204 and 1135 of the Water Resources Development Act.

Table 3. Marsh Creation Projects Constructed/Authorized under CWPPRA.

Project Name	Acres Benefited	Construction Completion Date
Bayou Labranche Wetland Creation	203	1994
Barataria Waterway Wetland Restoration	9	1996
West Belle Pass Headland Restoration	474	1998
Lake Chapeau Sediment Input and Hydrologic Restoration, Point Au Fer Island	509	1999
Sabine Refuge Marsh Creation	993	Cycles 1, 2, and 3 completed. Cycles 4 and 5 are pending.
Little Lake Shoreline Protection/Dedicated Dredging near Round Lake	713	2006
Goose Point/Point Platte Marsh Creation	436	2008
North Lake Mechant Landbridge Restoration	604	2009
Bayou Dupont Sediment Delivery System	326	2010
Dedicated Dredging on the Barataria Basin Landbridge	242	2010
West Lake Boudreaux Shoreline Protection and Marsh Creation	277	2011
East Marsh Island Marsh Creation	169	2011

Scientific studies in coastal Louisiana also provide support for the use of dedicated dredging to restore coastal wetlands. Most research conducted on dedicated dredging projects in coastal Louisiana has occurred in saline marsh habitats. Although the project area supports an a brackish marsh community, the response should be somewhat similar to that observed in saline marsh. Marshes created at the correct elevation take only a few years to develop vegetative communities similar to those in natural marshes (Edwards and Proffitt 2003). Percent vegetative cover also equals that found in natural marshes, but only after several years of growth (Proffitt and Young 1999). However, soil characteristics between created and natural marshes are often very different, with created marshes being lower in organic matter and higher in bulk density (Edwards and Proffitt 2003).

Thin-layer sediment deposition to the marsh surface (i.e., marsh nourishment) has also been investigated as a restoration technique in coastal Louisiana. Mendelssohn and Kuhn (1999) studied the impacts of sediment addition to a deteriorating saline marsh dominated by smooth cordgrass. Sediment addition ranging from trace amounts to nearly 24 inches above natural marsh elevations produced increases in plant cover and plant height. Sediment addition reduced flooding, allowed for better soil aeration, and lowered concentrations of phytotoxins which provided better conditions for plant growth. Ford et al. (1999) investigated the effects of thin-layer deposition of dredged material via spray dredging in a deteriorated saline marsh. One year following the addition of approximately 9 inches of sediment, percent cover of smooth cordgrass increased three-fold over pre-project conditions with no lasting negative impacts on the native marsh plant community.

The Preferred Alternative is supported by the LCWCRTF, which approved funding for engineering and design at their February 8, 2006, meeting and subsequently approved funding for construction at their January 21, 2009, meeting. The Preferred Alternative would create emergent marsh in the project area, increase its habitat value for fish and wildlife resources, and result in a net gain of 447 acres of marsh at the end of the project life compared to the No Action Alternative. The Preferred Alternative also supports the restoration strategies recommended for this region in the Coast 2050 Plan.

SECTION 6.0 COMPATIBILITY WITH CWPPRA AND COMMUNITY OBJECTIVES

The Preferred Alternative would help to achieve CWPPRA objectives for protection and restoration of Louisiana's coastal wetlands. The cumulative impact of all CWPPRA projects approved to date would result in the protection/creation/restoration of over 111,000 acres of coastal wetlands. Cumulative impacts of the CWPPRA Program are addressed in the Louisiana Coastal Wetlands Restoration Plan Main Report and Environmental Impact Statement (LCWCRTF 1993).

Community objectives would likely be enhanced by the proposed project. Common socioeconomic goals include the conservation of sustainable fishing, shrimping, crabbing, and hunting opportunities in the region. The general public also supports wetland restoration and preservation for fish and wildlife habitat, and for recreational, aesthetic, and other non-consumptive uses. In addition, the public is now much more aware of the surge reduction benefits provided by wetlands since the passage of Hurricanes Katrina and Rita in 2005.

SECTION 7.0   COMPLIANCE WITH LAWS, REGULATIONS AND POLICIES

This Environmental Assessment was prepared in compliance with the National Environmental Policy Act of 1969 (NEPA). It is consistent with the NEPA-compliance procedures contained in the Fish and Wildlife Service Manual (550 FW 1-3), and employs a systematic, interdisciplinary approach. The proposed action alternative involves disposal of fill material into waters or wetlands; therefore, an evaluation under Section 404(b)(1) of the Clean Water Act of 1977, as amended, is required, as well as State of Louisiana water quality certification under Section 401. A Section 404 permit (dated June 3, 2009) has been received from the U.S. Army Corps of Engineers as well as Water Quality Certification (dated March 31, 2009) from the Louisiana Department of Environmental Quality . In addition, the Louisiana Department of Natural Resources has determined that the project is consistent with the Louisiana Coastal Resources Program.

Under the MSFCMA, the Service initiated consultation with the National Marine Fisheries Service upon submission of a draft Environmental Assessment, and has evaluated project-related impacts to EFH within the project area. The Preferred Alternative would result in adverse impacts to some categories (i.e., mud bottom and estuarine water column) of EFH; however, more productive categories of EFH, such as estuarine emergent wetlands, would be created. Therefore, the Service finds that the Preferred Alternative would not result in net adverse impacts to habitats designated as EFH under the MSFCMA.

By correspondence dated October 6, 2008, the Louisiana Department of Culture, Recreation and Tourism indicated that they have no objection to implementation of the Preferred Alternative. No archaeological sites are located within the project area.

Pursuant to Executive Order 12898 (Environmental Justice for Minority Populations), the Service has determined that the Preferred Alternative would not result in disproportionately high and adverse human health or environmental impacts on minority and low-income populations.

The proposed action has been internally reviewed by the U.S. Fish and Wildlife Service for compliance with the Endangered Species Act of 1973, as amended. In addition, the proposed action has been reviewed for compliance with the Archeological and Historic Preservation Act of 1974; Executive Order 11988 (Floodplain Management); Executive Order 11990 (Protection of Wetlands); and Executive Order 13186 (Responsibilities of Federal Agencies to Protect Migratory Birds).

SECTION 8.0   PREPARER

This Environmental Assessment was prepared by Kevin J. Roy, Senior Field Biologist with the U.S. Fish and Wildlife Service, Lafayette Field Office, Lafayette, Louisiana.

SECTION 9.0   LITERATURE CITED

Barras, J.A., P.E. Bourgeois, and L.R. Handley. 1994. Land loss in coastal Louisiana 1956-90. National Biological Survey, National Wetlands Research Center Open Report 94-01. 4 pp.

Barras, J., Beville, S., Britsch, D., Hartley, S., Hawes, S., Johnston, J., Kemp, P., Kinler, Q., Martucci, A., Porthouse, J., Reed, D., Roy, K., Sapkota, S., and J. Suhayda. 2003. Historical and projected coastal Louisiana land changes: 1978-2050: USGS Open File Report 03-334, 39 pp.

Barras, J.A., Bernier, J.C., and Morton, R.A. 2008. Land area change in coastal Louisiana—a multidecadal perspective (from 1956 to 2006); land changes: 1978-2050: USGS Open File Report 03-334, 39 pp.

Boshart, W. 2003. BA-04 West Pointe a la Hache Siphon Construction Summary Data and Graphics. Louisiana Department of Natural Resources, Coastal Restoration Division, Baton Rouge, LA. 56 pp.

Castellanos, D.L. and L. P. Rozas. 2001. Nekton use of submerged aquatic vegetation, marsh, and shallow unvegetated bottom in the Atchafalaya River delta, a Louisiana tidal freshwater ecosystem. Estuaries. Vol. 24, No. 2, p. 184-197.

Chabreck, R. and G. Linscombe. 1978. Vegetative type map of the Louisiana coastal marshes. Louisiana Department of Wildlife and Fisheries, New Orleans.

Chabreck, R. and G. Linscombe. 1988. Vegetative type map of the Louisiana coastal marshes. Louisiana Department of Wildlife and Fisheries, Baton Rouge.

Chabreck, R.H., and G. Linscombe. 1997. Vegetative type map of the Louisiana coastal marshes: Baton Rouge, Louisiana Department of Wildlife and Fisheries.

Chabreck, R.H., Palmisano, A.W., Jr., and T. Joanen. 1968. Vegetative type map of the Louisiana coastal marshes: Baton Rouge, Louisiana Department of Wildlife and Fisheries.

Dahl, T.E. 2000. Status and trends of wetlands in the conterminous United States 1986 to 1997. U.S. Department of the Interior, Fish and Wildlife Service, Washington, D.C. 82 pp.

Dunbar, J.B., L.D. Britsch and E.B. Kemp, III. 1992. Land loss rates, report 3, Louisiana coastal plain. Technical Report GL-90-2. Vicksburg, MS.: U.S. Army Corps of Engineers, U.S. Waterways Experiment Station.

Edwards, K. R. and C. E. Proffitt. 2003. Comparison of wetland structural characteristics between created and natural salt marshes in southwest Louisiana, USA. Wetlands, Vol. 23, No. 2 pp. 344-356.

Eustis Engineering Services, L.L.C. 2007. Geotechnical Investigation, State of Louisiana, Lake Hermitage Marsh Creation (BA-42), Plaquemines Parish, LA. Metairie, LA.

Ford, M. A., D. R. Cahoon and J. C. Lynch. 1999. Restoring marsh elevation in a rapidly subsiding salt marsh by thin-layer deposition of dredged material. Ecological Engineering, Vol. 12, pp. 189-205.

Lindquist, D.C. 2008. Ecological Review: Lake Hermitage Marsh Creation. Office of Coastal Protection and Restoration. Baton Rouge, LA. 16 pp.

Lindquist, D.C. and S. R. Martin. 2007. Coastal restoration annual project reviews: December 2007. Louisiana Department of Natural Resources, Baton Rouge, LA. 123 pp.

Linscombe, G. and R.H. Chabreck. 2001. Vegetative type map of the Louisiana coastal marshes. Louisiana Department of Wildlife and Fisheries, Baton Rouge.

Louisiana Coastal Wetlands Conservation and Restoration Task Force. 1993. Louisiana coastal wetlands restoration plan, main report and environmental impact statement.

Louisiana Coastal Wetlands Conservation and Restoration Task Force and the Wetlands Conservation and Restoration Authority. 1998a. Coast 2050: toward a sustainable coastal Louisiana. Louisiana Department of Natural Resources. Baton Rouge, LA. 161 pp.

Louisiana Coastal Wetlands Conservation and Restoration Task Force and the Wetlands Conservation and Restoration Authority. 1998b. Coast 2050: toward a sustainable coastal Louisiana. Appendix D. Louisiana Department of Natural Resources. Baton Rouge, La. 170 pp.

Louisiana Department of Environmental Quality. 2006. Louisiana water quality inventory: integrated report. Baton Rouge, LA. 102 pp plus appendices.

Louisiana Fur and Alligator Advisory Council. 1997. 1996-97 annual report, Fur and Alligator Advisory Council. Louisiana Department of Wildlife and Fisheries. 23 pp. plus appendices.

Louisiana Office of Coastal Protection and Restoration. 2008. Lake Hermitage Marsh Creation Project (BA-42): Final (95%) Design Report. Baton Rouge, LA. 37 pp. plus appendices.

Mendelssohn, I. A. and N. L. Kuhn. 1999. The effects of sediment addition on salt mash vegetation and soil physico-chemistry. Pages 55-61 in L. P. Rozas, J. A. Nyman, C. E. Proffitt, N. N. Rabalais, D. J. Reed, and R. E. Turner (eds.), Recent Research in Coastal Louisiana: Natural System Function and Response to Human Influence. Louisiana Sea Grant College Program, Baton Rouge, LA.

Minello, T. J. and L. P. Rozas. 2002. Nekton in gulf coast wetlands: fine-scale distributions, landscape patterns, and restoration implications. Ecological Applications, 12(2), pp. 441-455.

O'Neil, T. 1949. Map of the southern part of Louisiana showing vegetation types of the Louisiana marshes.

Proffitt, C. E. and J. Young. 1999. Salt marsh plant colonization, growth, and dominance on large mudflats created using dredged sediments. Pages 218-228 *in* L. P. Rozas, J. A. Nyman, C. E. Proffitt, N. N. Rabalais, D. J. Reed, and R. E. Turner (eds.), Recent Research in Coastal Louisiana: Natural System Function and Response to Human Influence. Louisiana Sea Grant College Program, Baton Rouge, LA.

Reed, D. J., Ed. 1995. Status and Historical Trends of Hydrologic Modification, Reduction in Sediment Availability, and Habitat Loss/Modification in the Barataria and Terrebonne Estuarine System. BTNEP Publ. No. 20, Barataria-Terrebonne National Estuary Program, Thibodaux, Louisiana, 338 pp. plus Appendices.

Rozas, L. P. and T. J. Minello. 2001. Marsh terracing as a wetland restoration tool for creating fishery habitat. Wetlands. Vol. 21, No. 3, pp. 327-341.

Sasser, C.E., Visser, J.M., Mouton, E., Linscombe, J., and S.B. Hartley. 2007. Vegetation types in coastal Louisiana in 2007: U.S. Geological Survey Open-File Report 2008–1224.

Sigma Consulting Group, Inc. 2007. Survey Methodology Report, Lake Hermitage Marsh Creation (BA-42), Topographic, Bathymetric, and Magnetometer Survey. Baton Rouge, LA.

Southwick Associates. 2005. The economic benefits of fisheries, wildlife, and boat resources in the State of Louisiana. Louisiana Department of Wildlife and Fisheries. 22 pp. plus appendices.

Turner, R.E., and D.R. Cahoon, eds. 1987. Causes of wetland loss in the coastal central Gulf of Mexico. Volume II: Technical Narrative. Final report submitted to Mineral Management Service, New Orleans, Louisiana. Contract No. 14-12-0001-30252. OCS Study/MMS 87-0120. 400 pp.

Turner, R.E. 1990. Landscape development and coastal wetland losses in the northern Gulf of Mexico. Amer. Zool. 30:89-105.

U.S. Department of Commerce, National Oceanic and Atmospheric Administration, and National Marine Fisheries Service. 2001. Fisheries of the United States, 2001. Washington, D.C.

U.S Fish and Wildlife Service. 2008. Lake Hermitage Marsh Creation: project information sheet for wetland value assessment. 13 pp.

APPENDIX A - Detailed Drawings of Project Features

STATE OF LOUISIANA
COASTAL PROTECTION AND RESTORATION AUTHORITY

LAKE HERMITAGE
MARSH CREATION
BA-42
PLAQUEMINES PARISH

PRELIMINARY

THIS DOCUMENT IS INTENDED FOR REVIEW PURPOSES
ONLY. IT SHOULD NOT BE USED FOR CONSTRUCTION,
BIDDING, RECORDATION, CONVEYANCE, SALES, OR AS
THE BASIS FOR ISSUANCE OF A PERMIT.

Rudolph A. Simoneaux III, P.E.
Louisiana License Number 34747

LICENSURE CLASSIFICATION REQUIREMENTS:
MAJOR CATEGORY: HEAVY CONSTRUCTION
SUBCLASSIFICATION: DREDGING

STATE OF LOUISIANA
INSET MAP

PLAQUEMINES PARISH

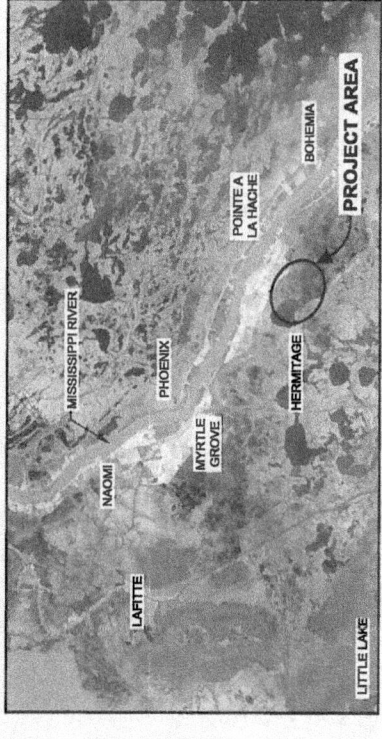

MISSISSIPPI RIVER

LAFITTE

NAOMI

PHOENIX

MYRTLE GROVE

POINTE A LA HACHE

HERMITAGE

BOHEMIA

PROJECT AREA

LITTLE LAKE

30,000'   15,000'   0'   30,000'   60,000'

CPRA
Louisiana Coastal Protection
and Restoration Authority

COASTAL PROTECTION AND
RESTORATION AUTHORITY
450 LAUREL STREET
BATON ROUGE, LOUISIANA 70801

DRAWN BY: KRISTI CANTU	DESIGNED BY: RUDOLPH SIMONEAUX, P.E.
REV.	DATE	DESCRIPTION	BY

CHIEF - RESTORATION DIVISION _____

ENGINEER MANAGER _____

PROJECT ENGINEER _____

LAKE HERMITAGE
MARSH CREATION

STATE PROJECT NUMBER: BA-42	TITLE SHEET
FEDERAL PROJECT NUMBER:	
APPROVED BY: JERRY CARROLL, P.E.	DATE: SEPTEMBER 2011	SHEET 1 OF 28

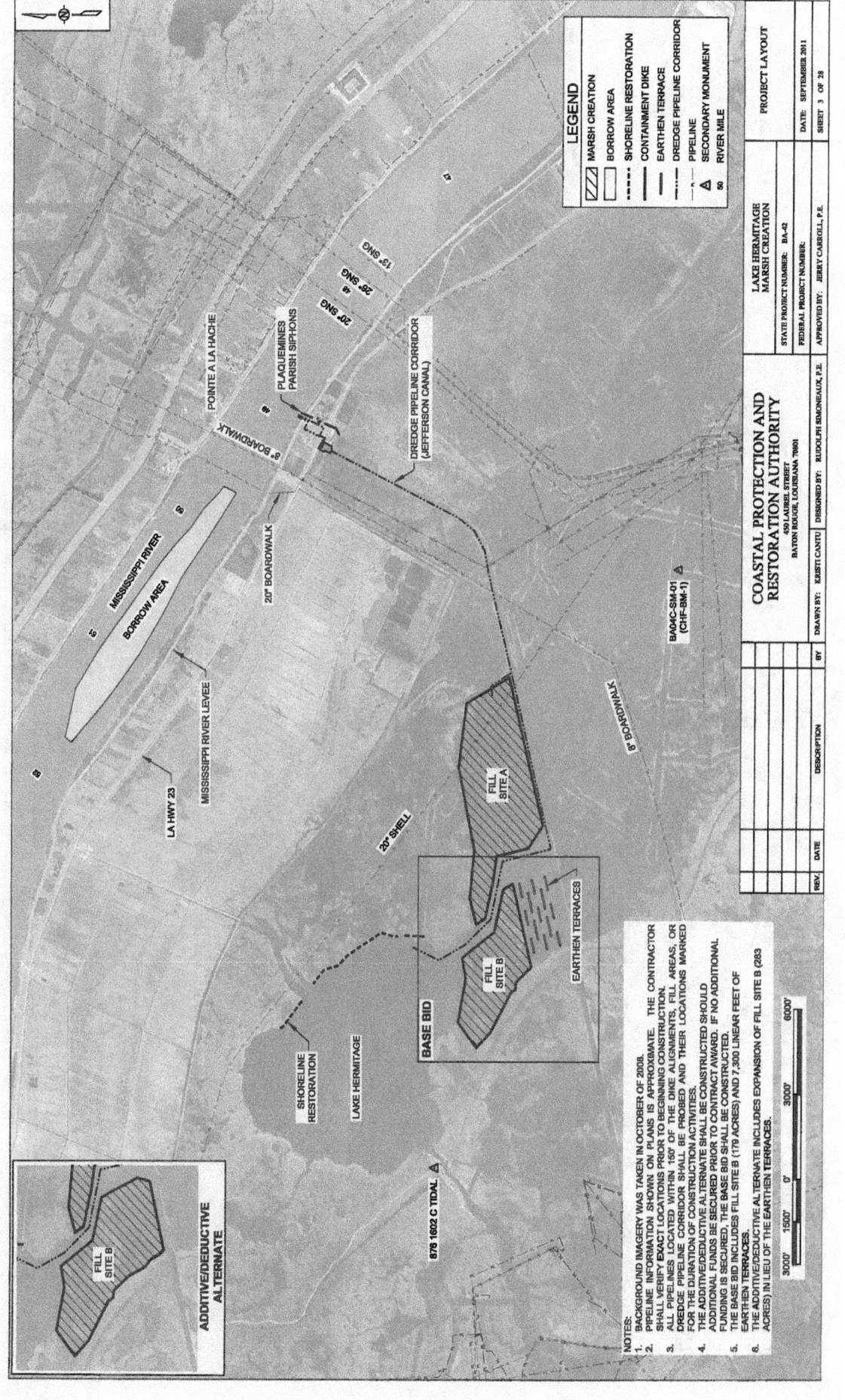

LEGEND

▨	MARSH CREATION
☐	BORROW AREA
▬▬▬	SHORELINE RESTORATION
▬▬▬	CONTAINMENT DIKE
▬▬▬	EARTHEN TERRACE
▬·▬	DREDGE PIPELINE CORRIDOR
▬	PIPELINE
△	SECONDARY MONUMENT
50	RIVER MILE

PROJECT LAYOUT

COASTAL PROTECTION AND
RESTORATION AUTHORITY
450 LAUREL STREET
BATON ROUGE, LOUISIANA 70801

LAKE HERMITAGE
MARSH CREATION

STATE PROJECT NUMBER: BA-42
FEDERAL PROJECT NUMBER:

DRAWN BY: KRISTI CANTU   DESIGNED BY: RUDOLPH SIMONEAUX, P.E.   APPROVED BY: JERRY CARROLL, P.E.

DATE: SEPTEMBER 2011

SHEET 3 OF 29

REV.	DATE	DESCRIPTION	BY

NOTES:

1. BACKGROUND IMAGERY WAS TAKEN IN OCTOBER OF 2008.
2. PIPELINE INFORMATION SHOWN ON PLANS IS APPROXIMATE. THE CONTRACTOR SHALL VERIFY EXACT LOCATIONS PRIOR TO BEGINNING CONSTRUCTION.
3. ALL PIPELINES LOCATED WITHIN 150' OF THE DIKE ALIGNMENTS, FILL AREAS, OR DREDGE PIPELINE CORRIDOR SHALL BE PROBED AND THEIR LOCATIONS MARKED FOR THE DURATION OF CONSTRUCTION ACTIVITIES.
4. THE ADDITIVE/DEDUCTIVE ALTERNATE SHALL BE CONSTRUCTED SHOULD ADDITIONAL FUNDS BE SECURED PRIOR TO CONTRACT AWARD. IF NO ADDITIONAL FUNDING IS SECURED, THE BASE BID SHALL BE CONSTRUCTED.
5. THE BASE BID INCLUDES FILL SITE B (179 ACRES) AND 7,300 LINEAR FEET OF EARTHEN TERRACES.
6. THE ADDITIVE/DEDUCTIVE ALTERNATE INCLUDES EXPANSION OF FILL SITE B (283 ACRES) IN LIEU OF THE EARTHEN TERRACES.

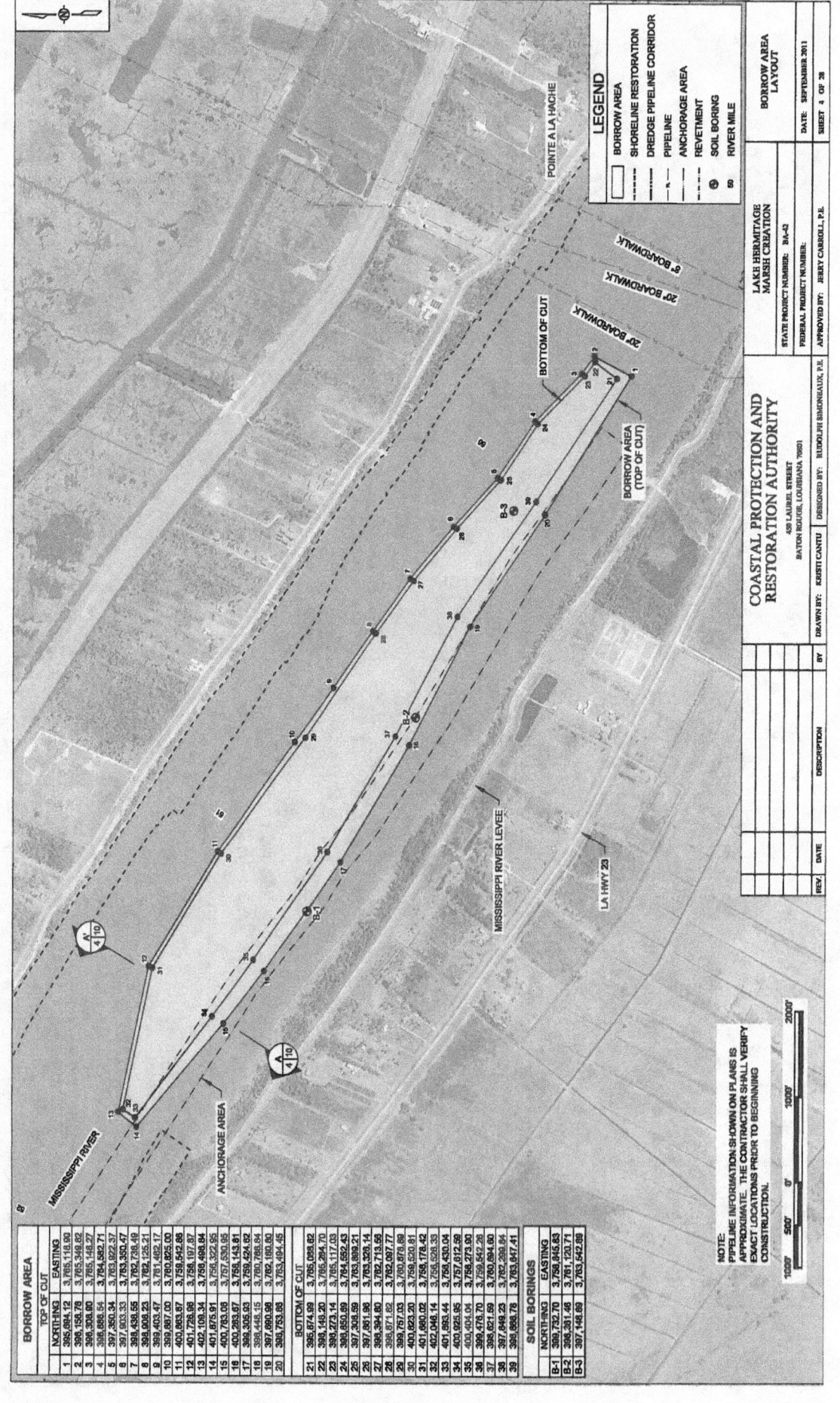

BORROW AREA

TOP OF CUT

	NORTHING	EASTING
1	395,694.12	3,765,116.90
2	396,156.78	3,765,349.82
3	396,303.00	3,765,146.27
4	396,688.54	3,764,562.71
5	397,350.34	3,763,922.37
6	397,203.33	3,763,350.47
7	394,436.55	3,762,738.49
8	394,906.23	3,762,125.21
9	399,403.47	3,761,492.17
10	399,897.00	3,760,925.00
11	400,683.87	3,759,542.86
12	401,726.96	3,758,197.87
13	402,109.34	3,758,498.84
14	401,875.91	3,758,322.95
15	400,783.06	3,757,530.66
16	400,283.67	3,758,143.81
17	399,305.93	3,759,424.62
18	399,448.15	3,760,786.84
19	397,680.98	3,762,160.80
20	396,753.66	3,763,494.45

BOTTOM OF CUT

	NORTHING	EASTING
21	395,674.99	3,765,093.82
22	396,149.20	3,765,284.70
23	396,273.14	3,765,117.03
24	396,650.99	3,764,552.43
25	397,308.59	3,763,909.21
26	397,081.96	3,763,329.14
27	398,394.80	3,762,713.56
28	398,871.62	3,762,097.77
29	399,757.03	3,760,876.69
30	400,622.20	3,759,520.61
31	401,680.02	3,758,176.42
32	402,048.14	3,758,508.33
33	401,893.44	3,758,430.04
34	400,925.95	3,757,612.59
35	400,404.04	3,758,273.00
36	399,478.70	3,759,542.26
37	398,621.59	3,760,894.60
38	397,649.23	3,762,299.84
39	396,688.76	3,763,547.08

SOIL BORINGS

	NORTHING	EASTING
B-1	390,732.70	3,758,845.63
B-2	398,361.46	3,761,120.71
B-3	397,146.69	3,763,542.08

NOTE:
PIPELINE INFORMATION SHOWN ON PLANS IS APPROXIMATE. THE CONTRACTOR SHALL VERIFY EXACT LOCATIONS PRIOR TO BEGINNING CONSTRUCTION.

1000' 500' 0' 1000' 2000'

LEGEND

☐	BORROW AREA
–·–	SHORELINE RESTORATION
––––	DREDGE PIPELINE CORRIDOR
—P—	PIPELINE
	ANCHORAGE AREA
–··–	REVETMENT
⊕	SOIL BORING
50	RIVER MILE

COASTAL PROTECTION AND RESTORATION AUTHORITY
439 LAUREL STREET
BATON ROUGE, LOUISIANA 70801

DRAWN BY: KRISTI CANTU	DESIGNED BY: RUDOLPH SIMONEAUX, P.E.	APPROVED BY: JERRY CARROLL, P.E.

REV.	DATE	DESCRIPTION	BY

LAKE HERMITAGE MARSH CREATION

STATE PROJECT NUMBER: BA-42

FEDERAL PROJECT NUMBER:

BORROW AREA LAYOUT

DATE: SEPTEMBER 2011

SHEET 4 OF 28

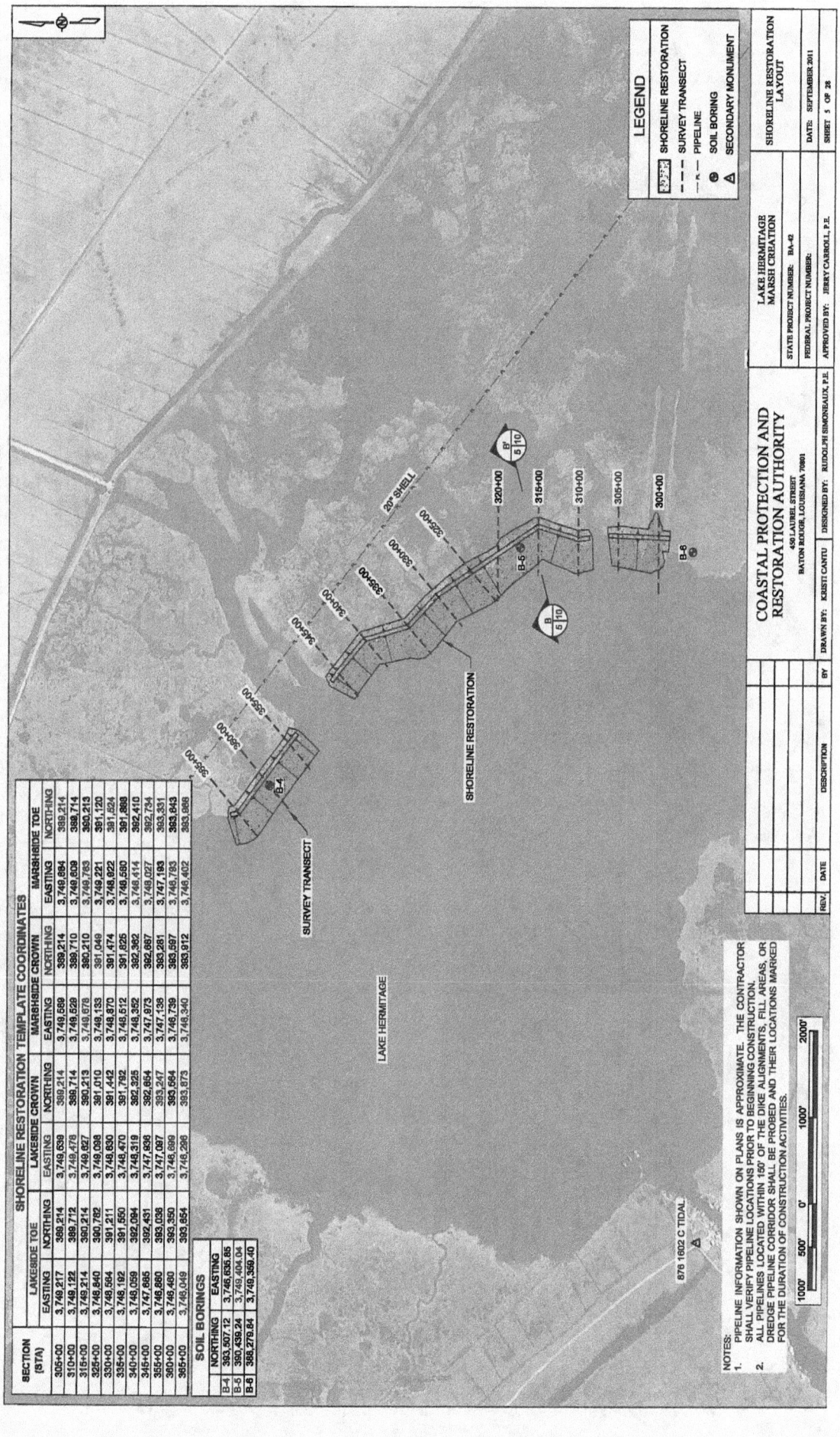

SHORELINE RESTORATION TEMPLATE COORDINATES

SECTION (STA)	LAKESIDE TOE EASTING	LAKESIDE TOE NORTHING	LAKESIDE CROWN EASTING	LAKESIDE CROWN NORTHING	MARSHSIDE CROWN EASTING	MARSHSIDE CROWN NORTHING	MARSHSIDE TOE EASTING	MARSHSIDE TOE NORTHING
305+00	3,749,217	389,214	3,749,639	389,214	3,749,589	389,214	3,749,694	389,214
310+00	3,749,192	389,712	3,749,478	389,714	3,749,529	389,710	3,749,609	389,714
315+00	3,749,214	390,214	3,749,627	389,710	3,749,678	390,210	3,749,763	390,213
325+00	3,748,840	390,782	3,749,098	391,010	3,749,133	391,049	3,749,221	391,120
330+00	3,748,564	391,211	3,748,630	391,442	3,748,870	391,474	3,748,822	381,624
335+00	3,748,192	391,550	3,748,470	391,792	3,748,612	391,826	3,748,580	391,866
340+00	3,748,059	392,094	3,748,319	392,325	3,748,352	392,382	3,748,614	392,410
345+00	3,747,665	392,431	3,747,898	392,664	3,747,973	392,687	3,748,027	392,734
355+00	3,746,680	393,036	3,747,087	393,247	3,747,138	393,281	3,747,193	393,331
360+00	3,746,480	393,350	3,746,699	393,564	3,746,739	393,597	3,746,783	393,643
365+00	3,746,049	393,654	3,746,296	393,873	3,746,340	393,912	3,746,602	393,966

SOIL BORINGS

	NORTHING	EASTING
B-4	393,507.12	3,746,636.65
B-5	390,439.24	3,749,404.04
B-6	386,279.84	3,749,360.40

LAKE HERMITAGE

SURVEY TRANSECT

SHORELINE RESTORATION

20" SHELL

876 1602 C TIDAL

LEGEND
SHORELINE RESTORATION
SURVEY TRANSECT
PIPELINE
SOIL BORING
SECONDARY MONUMENT

NOTES:
1. PIPELINE INFORMATION SHOWN ON PLANS IS APPROXIMATE. THE CONTRACTOR SHALL VERIFY PIPELINE LOCATIONS PRIOR TO BEGINNING CONSTRUCTION.
2. ALL PIPELINES LOCATED WITHIN 150' OF THE DIKE ALIGNMENTS, FILL AREAS, OR DREDGE PIPELINE CORRIDOR SHALL BE PROBED AND THEIR LOCATIONS MARKED FOR THE DURATION OF CONSTRUCTION ACTIVITIES.

1000' 500' 0' 1000' 2000'

COASTAL PROTECTION AND RESTORATION AUTHORITY
450 LAUREL STREET
BATON ROUGE, LOUISIANA 70801

LAKE HERMITAGE MARSH CREATION
STATE PROJECT NUMBER: BA-42
FEDERAL PROJECT NUMBER:
APPROVED BY: JERRY CARROLL, P.E.

DRAWN BY: KRISTI CANTU   DESIGNED BY: RUDOLPH SIMONEAUX, P.E.

SHORELINE RESTORATION LAYOUT
DATE: SEPTEMBER 2011   SHEET 5 OF 28

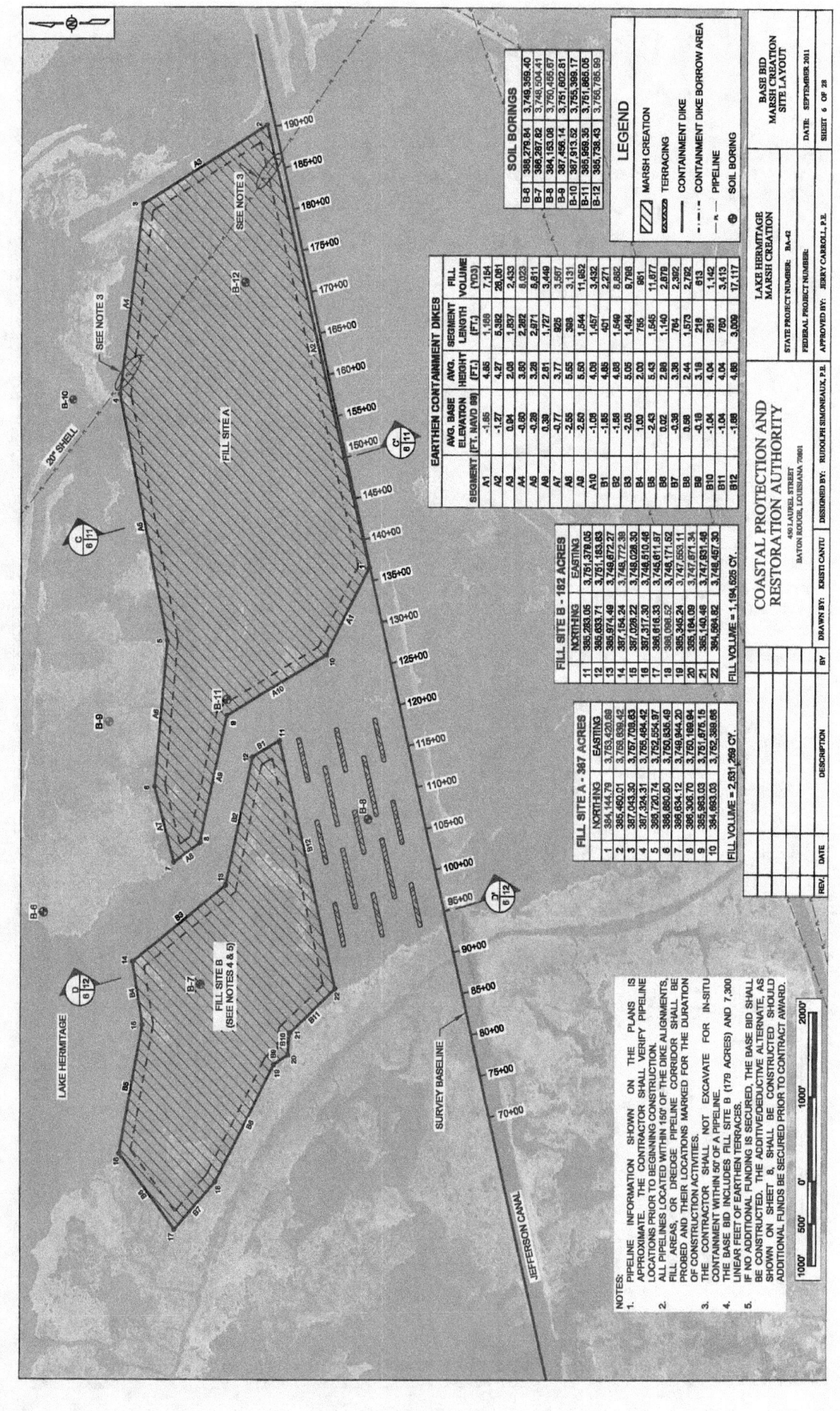

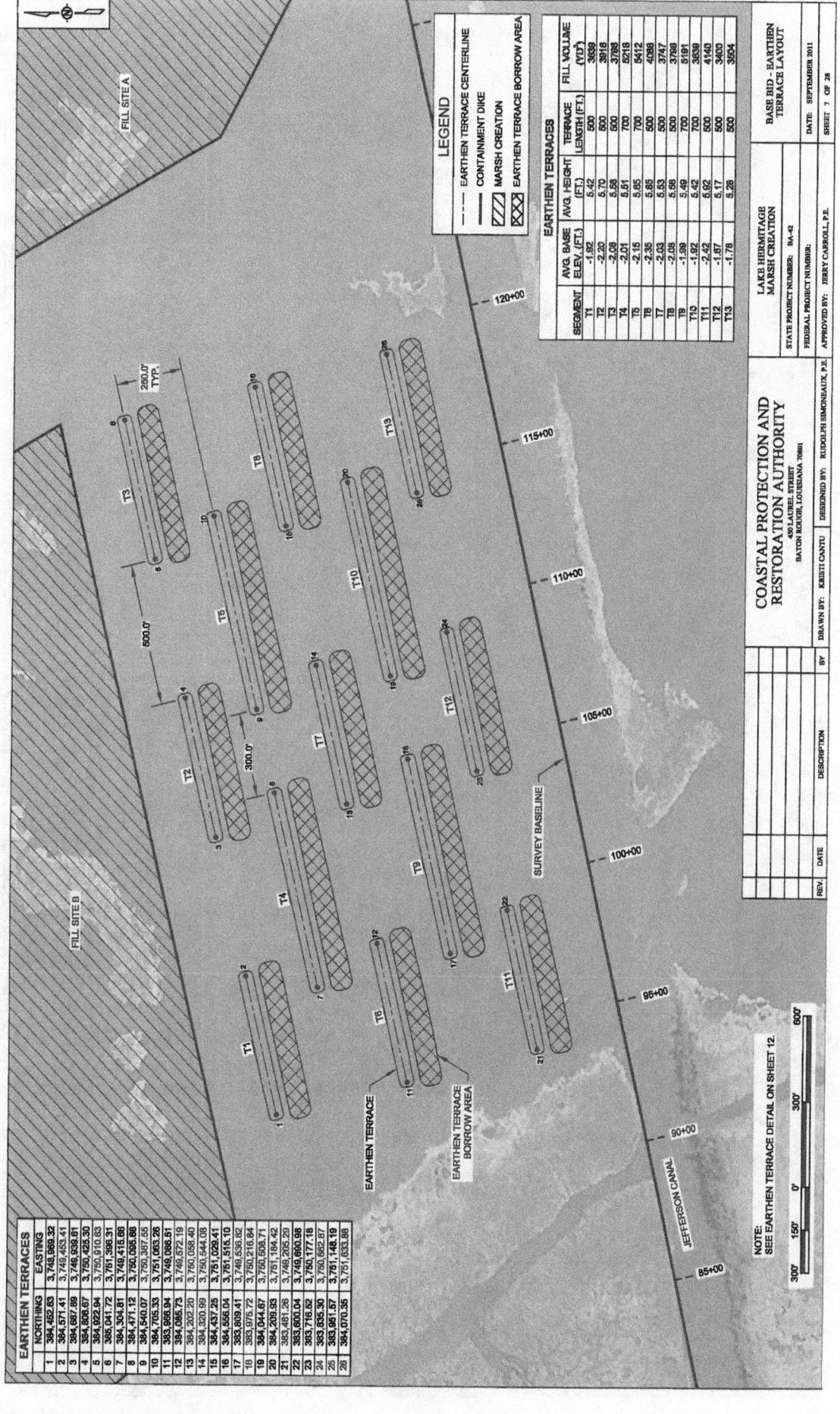

COASTAL PROTECTION AND
RESTORATION AUTHORITY
450 LAUREL STREET
BATON ROUGE, LOUISIANA 70801

LAKE HERMITAGE
MARSH CREATION

STATE PROJECT NUMBER: BA-42

FEDERAL PROJECT NUMBER:

BASE BID - EARTHEN
TERRACE LAYOUT

DATE: SEPTEMBER 2011

SHEET 7 OF 28

EARTHEN TERRACES

SEGMENT	AVG. BASE ELEV. (FT.)	AVG. HEIGHT (FT.)	TERRACE LENGTH (FT.)	FILL VOLUME (YD³)
T1	-1.92	5.42	500	3639
T2	-2.20	5.70	500	3916
T3	-2.08	5.58	500	3768
T4	-2.01	5.51	700	5218
T5	-2.35	5.85	700	5412
T6	-2.15	5.65	500	4098
T7	-2.03	5.53	500	3747
T8	-2.08	5.58	500	3798
T9	-1.99	5.49	700	5191
T10	-1.92	5.42	700	3639
T11	-2.42	5.92	500	4140
T12	-1.67	5.17	500	3400
T13	-1.78	5.28	500	3504

LEGEND

- — · — EARTHEN TERRACE CENTERLINE
- ———— CONTAINMENT DIKE
- //// MARSH CREATION
- ⊠⊠ EARTHEN TERRACE BORROW AREA

EARTHEN TERRACES

	NORTHING	EASTING
1	384,462.63	3,748,969.32
2	384,571.41	3,749,453.41
3	384,667.89	3,749,939.61
4	384,806.67	3,750,425.30
5	384,922.94	3,750,910.63
6	385,041.72	3,751,396.31
7	384,304.81	3,749,415.66
8	384,471.12	3,750,005.66
9	384,540.07	3,750,387.55
10	384,705.53	3,751,063.26
11	383,966.94	3,749,088.51
12	384,065.73	3,749,572.19
13	384,202.20	3,750,058.40
14	384,320.99	3,750,544.08
15	384,437.25	3,751,029.41
16	384,556.04	3,751,515.10
17	383,809.41	3,749,536.82
18	383,976.72	3,750,216.84
19	384,044.67	3,750,508.71
20	384,209.93	3,751,184.42
21	383,481.26	3,749,205.29
22	383,600.04	3,749,690.98
23	383,716.62	3,750,177.18
24	383,835.30	3,750,662.87
25	383,951.57	3,751,148.19
26	384,070.35	3,751,633.88

REV.	DATE		DESCRIPTION	BY

DRAWN BY: KRISTI CANTU   DESIGNED BY: RUDOLPH SIMONEAUX, P.E.   APPROVED BY: JERRY CARROLL, P.E.

FILL SITE A

FILL SITE B

EARTHEN TERRACE

EARTHEN TERRACE
BORROW AREA

SURVEY BASELINE

JEFFERSON CANAL

NOTE:
SEE EARTHEN TERRACE DETAIL ON SHEET 12.

300' 150' 0' 300' 600'

260.0' TYP.

500.0'

300.0'

T1 T2 T3 T4 T5 T6 T7 T8 T9 T10 T11 T12 T13

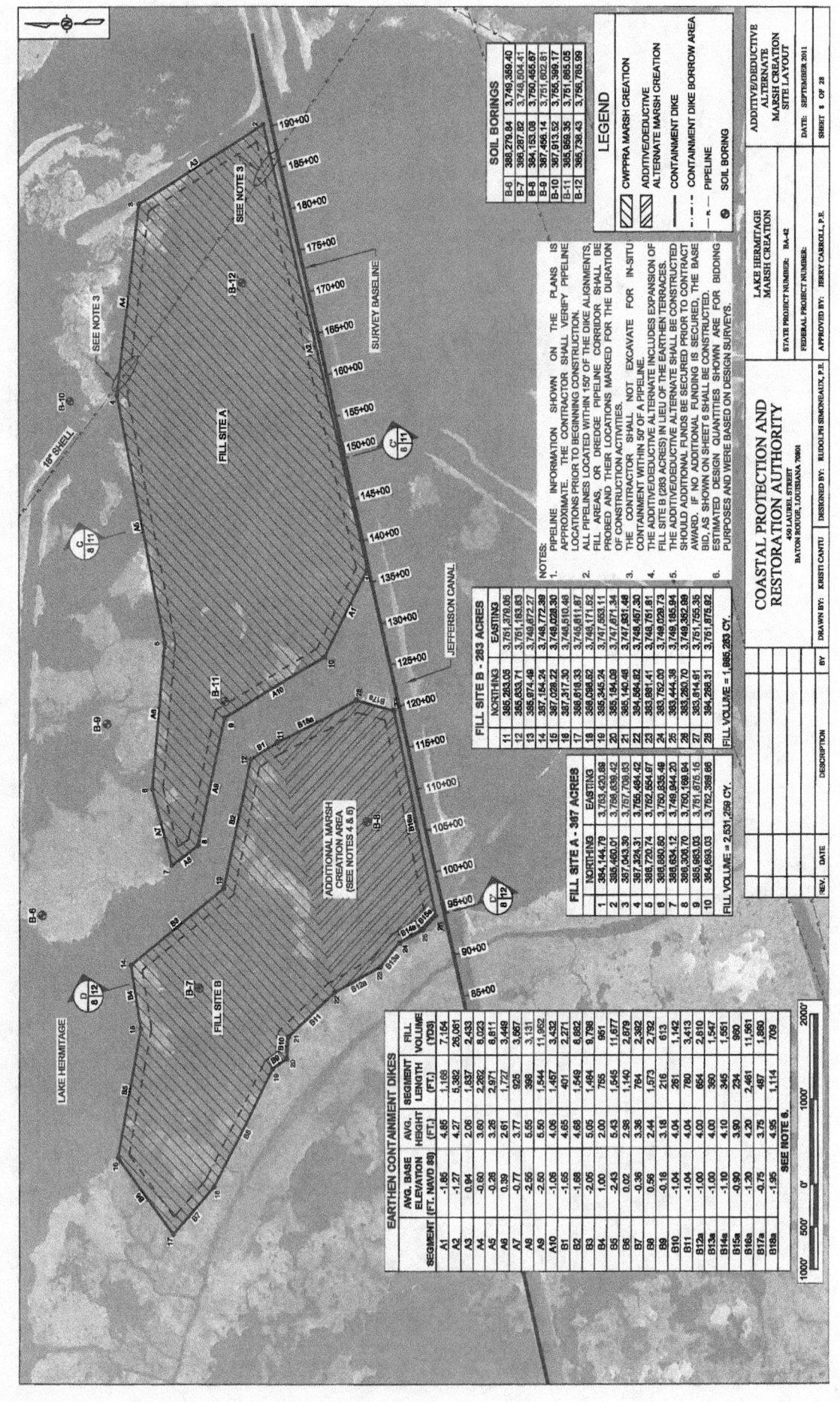

EARTHEN CONTAINMENT DIKES

SEGMENT	AVG. BASE ELEVATION (FT, NAVD 88)	AVG. HEIGHT (FT.)	SEGMENT LENGTH (FT.)	FILL VOLUME (YD3)
A1	-1.86	4.86	1,169	7,164
A3	-1.27	4.27	5,382	26,061
A4	0.94	2.06	1,837	2,433
A4	-0.60	3.60	2,262	8,023
A5	-0.26	3.26	2,971	8,611
A6	0.39	2.61	1,727	3,449
A7	-0.77	3.77	925	3,667
A9	-2.56	5.55	398	3,131
A9	-2.50	5.50	1,544	11,952
A10	-1.06	4.06	1,457	3,432
B1	-1.65	4.65	401	2,271
B2	-1.68	4.68	1,549	6,682
B3	-2.05	5.05	1,484	9,798
B4	1.00	2.00	765	961
B6	-2.43	5.43	1,545	11,677
B6	0.02	2.98	1,140	2,879
B7	-0.36	3.36	784	2,382
B8	0.56	2.44	1,573	2,792
B9	-0.18	3.18	216	613
B10	-1.04	4.04	261	1,142
B11	-1.04	4.04	780	3,413
B12a	-1.00	4.00	654	2,810
B13a	-1.00	4.00	360	1,547
B14a	-1.10	4.10	345	1,551
B15a	-0.90	3.90	234	960
B16a	-1.20	4.20	2,461	11,561
B17a	-0.75	3.75	487	1,860
B18a	-1.95	4.95	1,114	709

SEE NOTE 8.

FILL SITE A - 367 ACRES

	NORTHING	EASTING
1	384,144.79	3,753,420.69
2	385,460.01	3,756,639.42
3	387,043.30	3,757,708.03
4	387,324.31	3,759,484.42
5	388,720.74	3,752,554.97
6	388,680.60	3,750,535.46
7	386,834.12	3,749,944.20
8	386,306.70	3,750,169.64
9	385,863.03	3,751,675.15
10	384,693.03	3,752,389.66

FILL VOLUME = 2,531,259 CY.

FILL SITE B - 283 ACRES

	NORTHING	EASTING
11	385,263.05	3,751,379.05
12	385,633.71	3,751,183.63
13	385,974.49	3,749,672.27
14	387,154.24	3,749,772.39
15	387,028.22	3,749,028.30
16	387,029.22	3,749,028.30
17	387,317.30	3,748,510.46
18	386,618.33	3,745,611.87
19	386,098.62	3,747,563.11
20	385,345.24	3,747,671.34
21	385,184.09	3,747,931.48
22	385,140.46	3,749,457.30
23	394,564.82	3,748,751.81
24	383,881.41	3,748,029.73
25	383,752.00	3,749,185.94
26	383,444.39	3,749,382.99
27	383,814.91	3,751,755.35
28	384,266.31	3,751,675.92

FILL VOLUME = 1,895,263 CY.

SOIL BORINGS

B-6	388,279.84	3,749,369.40
B-7	386,287.82	3,748,504.41
B-8	384,153.08	3,750,465.67
B-9	387,456.14	3,751,602.81
B-10	387,913.52	3,755,399.17
B-11	385,959.35	3,751,685.05
B-12	365,738.43	3,756,785.99

LEGEND

- CWPPRA MARSH CREATION
- ADDITIVE/DEDUCTIVE ALTERNATE MARSH CREATION
- —— CONTAINMENT DIKE
- ---- CONTAINMENT DIKE BORROW AREA
- -..- PIPELINE
- ⊕ SOIL BORING

NOTES:

1. PIPELINE INFORMATION SHOWN ON THE PLANS IS APPROXIMATE. THE CONTRACTOR SHALL VERIFY PIPELINE LOCATIONS PRIOR TO BEGINNING CONSTRUCTION.

2. ALL PIPELINES LOCATED WITHIN 150' OF THE DIKE ALIGNMENTS, FILL AREAS, OR DREDGE PIPELINE CORRIDOR SHALL BE PROBED AND THEIR LOCATIONS MARKED FOR THE DURATION OF CONSTRUCTION ACTIVITIES.

3. THE CONTRACTOR SHALL NOT EXCAVATE FOR IN-SITU CONTAINMENT WITHIN 50' OF A PIPELINE.

4. THE ADDITIVE/DEDUCTIVE ALTERNATE INCLUDES EXPANSION OF FILL SITE B (283 ACRES) IN LIEU OF THE EARTHEN TERRACES.

5. THE ADDITIVE/DEDUCTIVE ALTERNATE SHALL BE CONSTRUCTED SHOULD ADDITIONAL FUNDS BE SECURED PRIOR TO CONTRACT AWARD. IF NO ADDITIONAL FUNDING IS SECURED, THE BASE BID, AS SHOWN ON SHEET 6 SHALL BE CONSTRUCTED.

6. ESTIMATED DESIGN QUANTITIES SHOWN ARE FOR BIDDING PURPOSES AND WERE BASED ON DESIGN SURVEYS.

COASTAL PROTECTION AND RESTORATION AUTHORITY
450 LAUREL STREET
BATON ROUGE, LOUISIANA 70801

DRAWN BY: KRISTI CANTU	DESIGNED BY: RUDOLPH SIMONEAUX, P.E.	APPROVED BY: JERRY CARROLL, P.E.

LAKE HERMITAGE MARSH CREATION

STATE PROJECT NUMBER: BA-42

FEDERAL PROJECT NUMBER:

ADDITIVE/DEDUCTIVE ALTERNATE MARSH CREATION SITE LAYOUT

DATE: SEPTEMBER 2011

SHEET 8 OF 28

REV.	DATE	DESCRIPTION	BY

LAKE HERMITAGE

15" SHELL

JEFFERSON CANAL

SURVEY BASELINE

FILL SITE A

FILL SITE B

ADDITIONAL MARSH CREATION AREA (SEE NOTES 4 & 6)

SEE NOTE 3

1000' 500' 0' 1000' 2000'

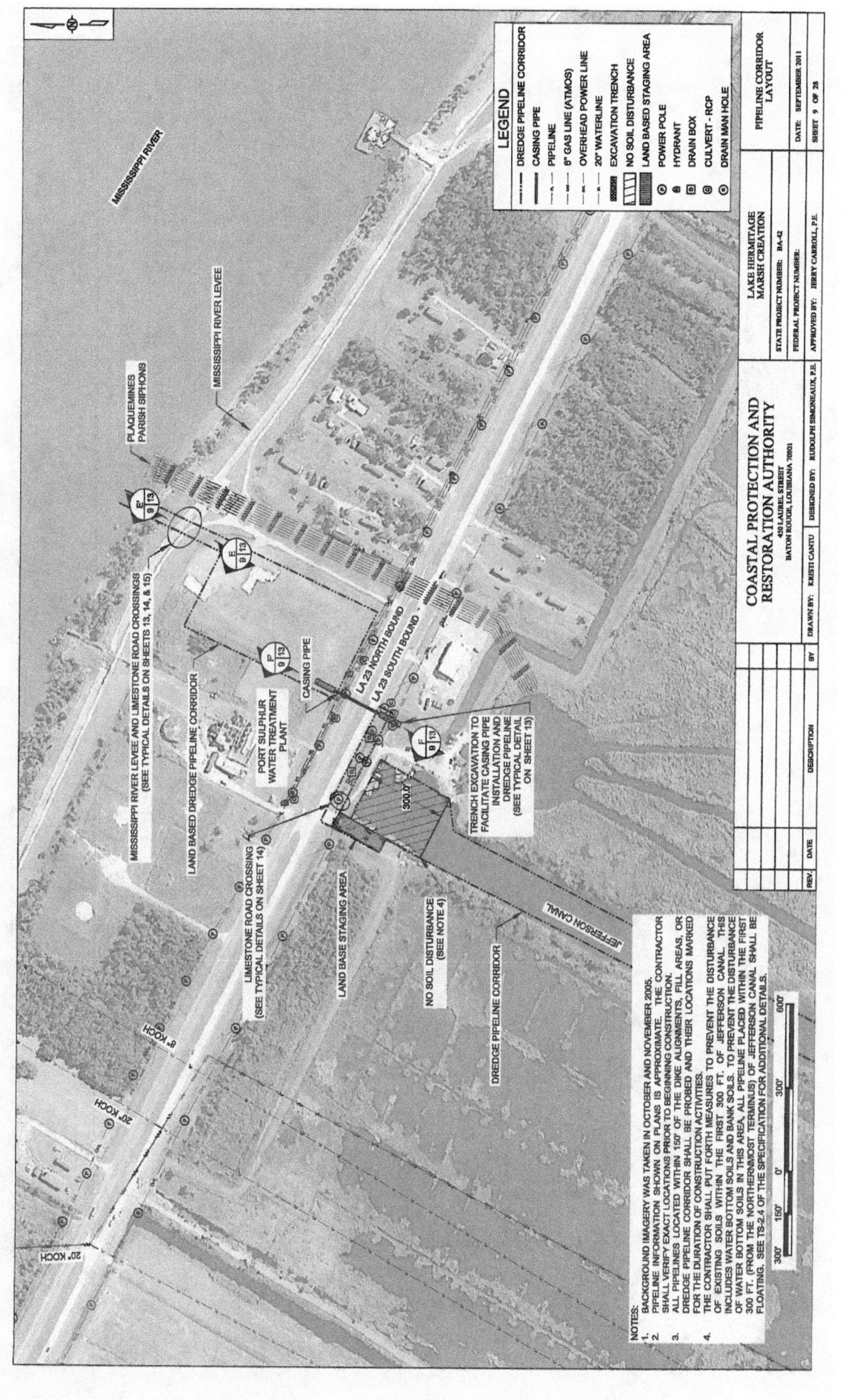

MISSISSIPPI RIVER

MISSISSIPPI RIVER LEVEE

PLAQUEMINES PARISH SIPHONS

MISSISSIPPI RIVER LEVEE AND LIMESTONE ROAD CROSSINGS
(SEE TYPICAL DETAILS ON SHEETS 13, 14, & 15)

LAND BASED DREDGE PIPELINE CORRIDOR

PORT SULPHUR WATER TREATMENT PLANT

CASING PIPE

LA 23 NORTH BOUND

LA 23 SOUTH BOUND

TRENCH EXCAVATION TO FACILITATE CASING PIPE INSTALLATION AND DREDGE PIPELINE (SEE TYPICAL DETAIL ON SHEET 13)

LIMESTONE ROAD CROSSING (SEE TYPICAL DETAILS ON SHEET 14)

LAND BASE STAGING AREA

300.0'

NO SOIL DISTURBANCE (SEE NOTE 4)

DREDGE PIPELINE CORRIDOR

JEFFERSON CANAL

8" KOCH
20" KOCH
20" KOCH

LEGEND

	DREDGE PIPELINE CORRIDOR
	CASING PIPE
	PIPELINE
	6" GAS LINE (ATMOS)
	OVERHEAD POWER LINE
	20" WATERLINE
	EXCAVATION TRENCH
	NO SOIL DISTURBANCE
	LAND BASED STAGING AREA
℗	POWER POLE
⬡	HYDRANT
⬛	DRAIN BOX
⊙	CULVERT - RCP
⊙	DRAIN MAN HOLE

NOTES:

1. BACKGROUND IMAGERY WAS TAKEN IN OCTOBER AND NOVEMBER 2005.
2. PIPELINE INFORMATION SHOWN ON PLANS IS APPROXIMATE. THE CONTRACTOR SHALL VERIFY EXACT LOCATIONS PRIOR TO BEGINNING CONSTRUCTION.
3. ALL PIPELINES LOCATED WITHIN 150' OF THE DIKE ALIGNMENTS, FILL AREAS, OR DREDGE PIPELINE CORRIDOR SHALL BE PROBED AND THEIR LOCATIONS MARKED FOR THE DURATION OF CONSTRUCTION ACTIVITIES.
4. THE CONTRACTOR SHALL PUT FORTH MEASURES TO PREVENT THE DISTURBANCE OF EXISTING SOILS WITHIN THE FIRST 300 FT. OF JEFFERSON CANAL. THIS INCLUDES WATER BOTTOM SOILS AND BANK SLS. TO PREVENT THE DISTURBANCE OF WATER BOTTOM SOILS IN THIS AREA, ALL PIPELINE PLACED WITHIN THE FIRST 300 FT. (FROM THE NORTHERNMOST TERMINUS) OF JEFFERSON CANAL SHALL BE FLOATING. SEE TS-2.4 OF THE SPECIFICATION FOR ADDITIONAL DETAILS.

COASTAL PROTECTION AND RESTORATION AUTHORITY
450 LAUREL STREET
BATON ROUGE, LOUISIANA 70801

| DRAWN BY: KRISTI CANTU | DESIGNED BY: RUDOLPH SIMONEAUX, P.E. |
| APPROVED BY: JERRY CARROLL, P.E. |

LAKE HERMITAGE
MARSH CREATION

STATE PROJECT NUMBER: BA-42

FEDERAL PROJECT NUMBER:

PIPELINE CORRIDOR LAYOUT

DATE: SEPTEMBER 2011

SHEET 9 OF 28

REV.	DATE	DESCRIPTION	BY

300' 150' 0' 300' 600'

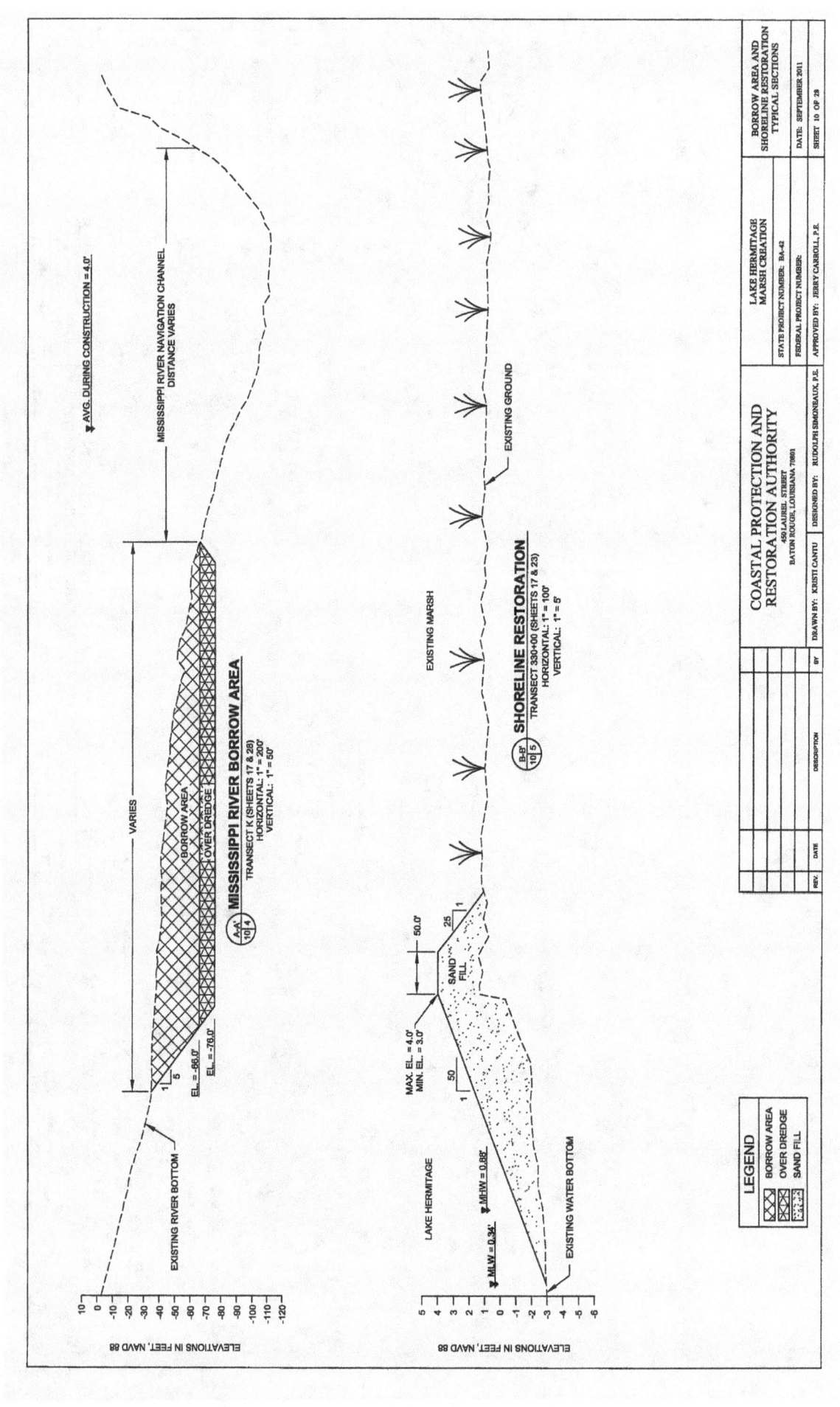

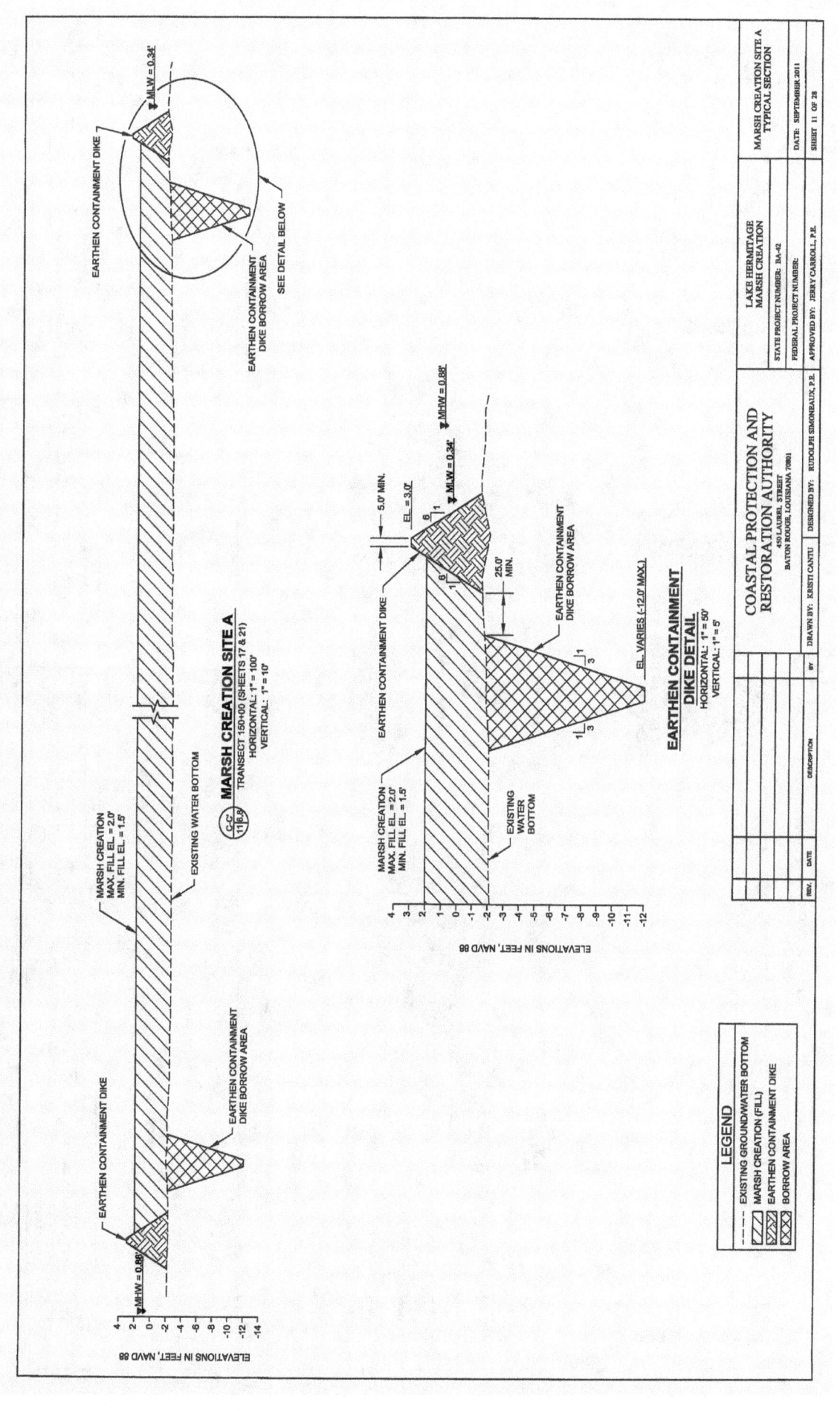

MARSH CREATION SITE A
TRANSECT 150+00 (SHEETS 17 & 21)
HORIZONTAL: 1" = 100'
VERTICAL: 1" = 10'

EARTHEN CONTAINMENT DIKE

MARSH CREATION
MAX. FILL EL = 2.0'
MIN. FILL EL = 1.5'

EXISTING WATER BOTTOM

EARTHEN CONTAINMENT
DIKE BORROW AREA

SEE DETAIL BELOW

MHW = 0.88'
MLW = 0.34'

EARTHEN CONTAINMENT
DIKE BORROW AREA

ELEVATIONS IN FEET, NAVD 88

EARTHEN CONTAINMENT
DIKE DETAIL
HORIZONTAL: 1" = 50'
VERTICAL: 1" = 5'

EARTHEN CONTAINMENT DIKE

5.0' MIN.

EL = 3.0'

MHW = 0.88'
MLW = 0.34'

MARSH CREATION
MAX. FILL EL = 2.0'
MIN. FILL EL = 1.5'

25.0'
MIN.

EXISTING
WATER
BOTTOM

EARTHEN CONTAINMENT
DIKE BORROW AREA

EL VARIES (-12.0' MAX.)

ELEVATIONS IN FEET, NAVD 88

LEGEND

EXISTING GROUNDWATER BOTTOM
MARSH CREATION (FILL)
EARTHEN CONTAINMENT DIKE
BORROW AREA

COASTAL PROTECTION AND
RESTORATION AUTHORITY
450 LAUREL STREET
BATON ROUGE, LOUISIANA 70801

LAKE HERMITAGE
MARSH CREATION
STATE PROJECT NUMBER: BA-42
FEDERAL PROJECT NUMBER:

MARSH CREATION SITE A
TYPICAL SECTION

DRAWN BY: KRISTI CANTU   DESIGNED BY: RUDOLPH SIMONEAUX, P.E.   APPROVED BY: JERRY CARROLL, P.E.

DATE: SEPTEMBER 2011

SHEET 11 OF 28

REV.   DATE   DESCRIPTION   BY

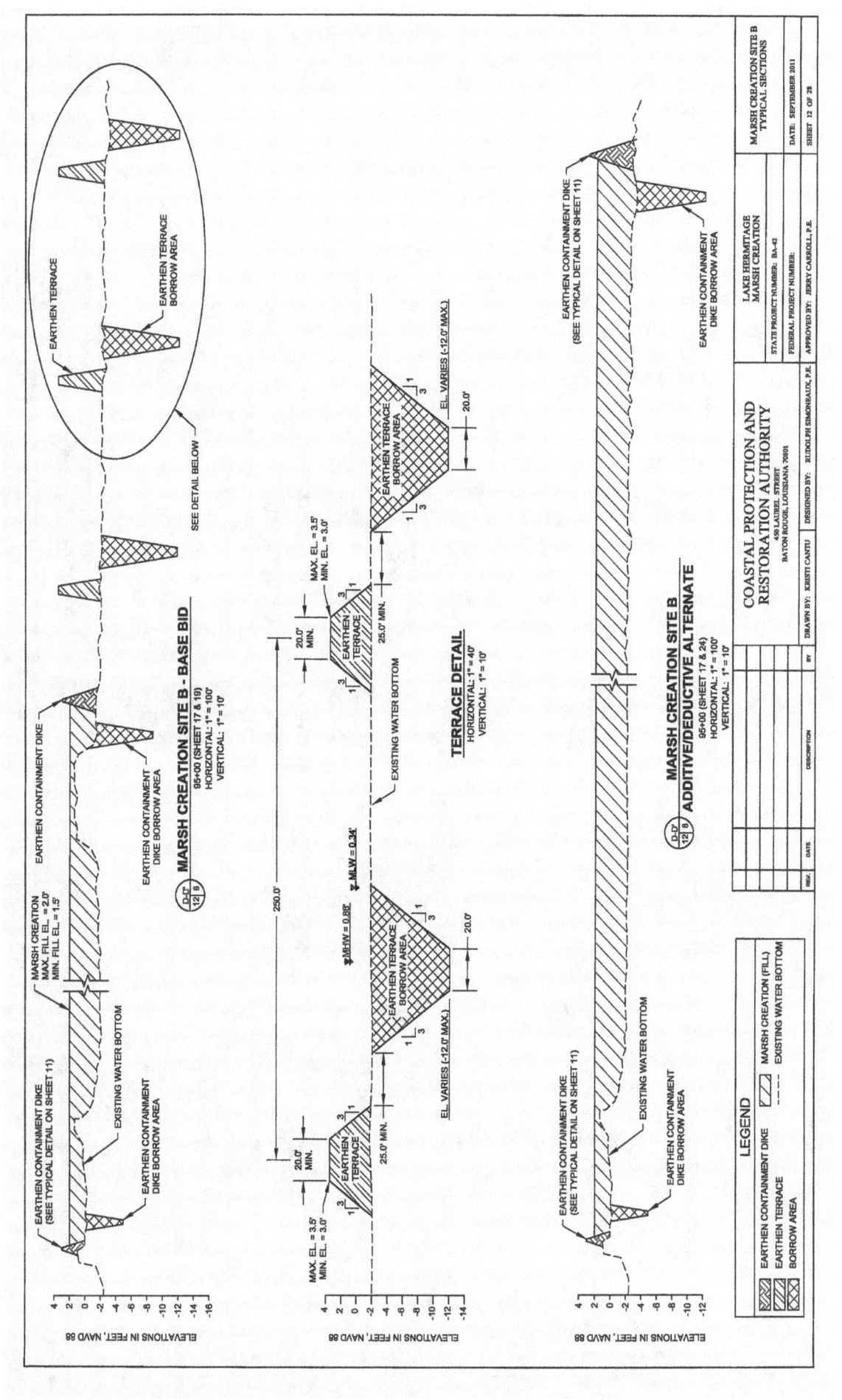

MARSH CREATION SITE B - BASE BID
95+00 (SHEET 17 & 19)
HORIZONTAL: 1" = 100'
VERTICAL: 1" = 10'

TERRACE DETAIL
HORIZONTAL: 1" = 40'
VERTICAL: 1" = 10'

MARSH CREATION SITE B
ADDITIVE/DEDUCTIVE ALTERNATE
95+00 (SHEET 17 & 24)
HORIZONTAL: 1" = 100'
VERTICAL: 1" = 10'

LEGEND

EARTHEN CONTAINMENT DIKE   MARSH CREATION (FILL)

EARTHEN TERRACE   EXISTING WATER BOTTOM

BORROW AREA

COASTAL PROTECTION AND
RESTORATION AUTHORITY
450 LAUREL STREET
BATON ROUGE, LOUISIANA 70801

LAKE HERMITAGE
MARSH CREATION

STATE PROJECT NUMBER: BA-42
FEDERAL PROJECT NUMBER:

MARSH CREATION SITE B
TYPICAL SECTIONS

DATE: SEPTEMBER 2011

SHEET 12 OF 26

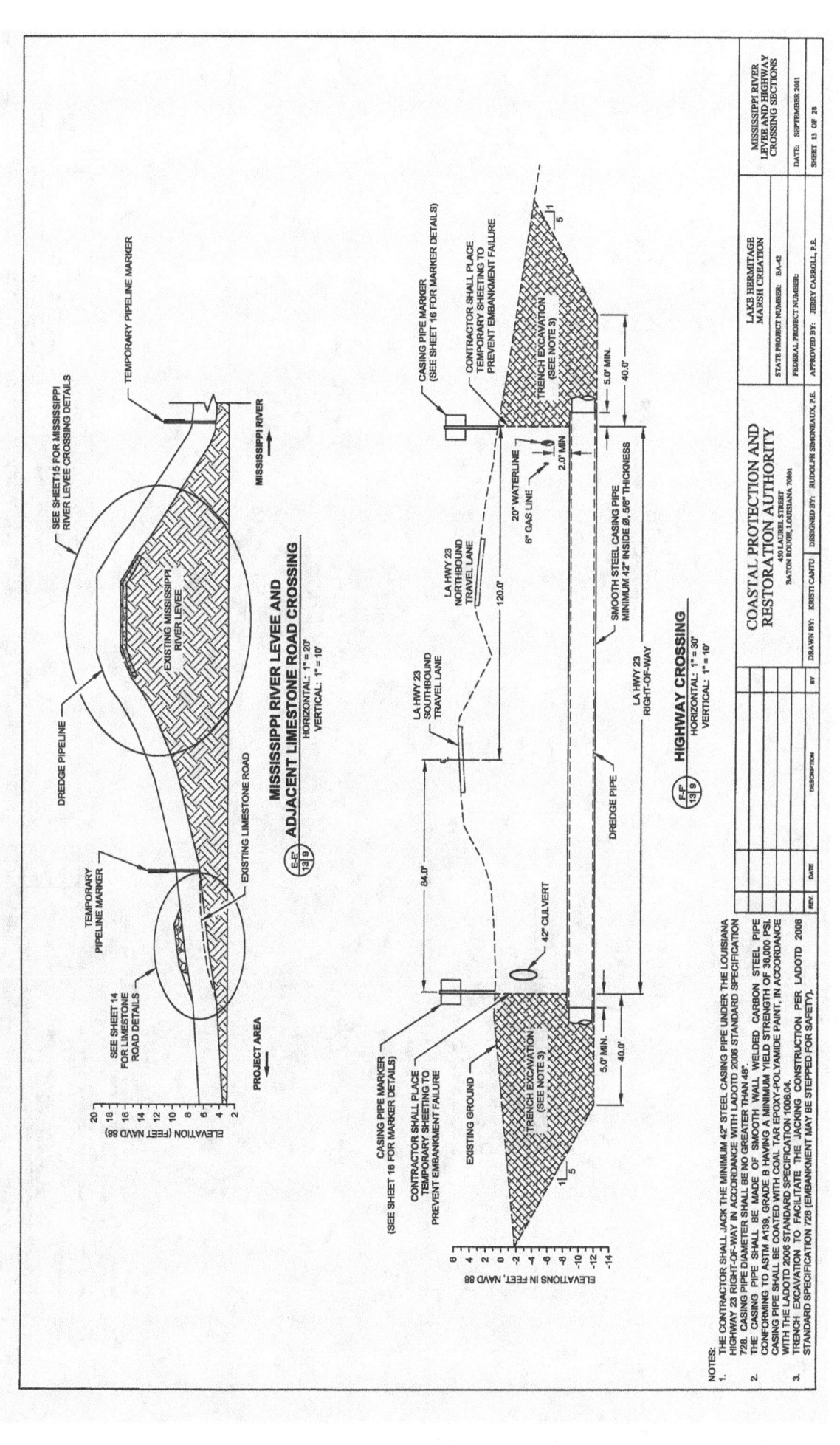

MISSISSIPPI RIVER LEVEE AND
ADJACENT LIMESTONE ROAD CROSSING
HORIZONTAL: 1" = 20'
VERTICAL: 1" = 10'

HIGHWAY CROSSING
HORIZONTAL: 1" = 30'
VERTICAL: 1" = 10'

NOTES:

1. THE CONTRACTOR SHALL JACK THE MINIMUM 42" STEEL CASING PIPE UNDER THE LOUISIANA HIGHWAY 23 RIGHT-OF-WAY IN ACCORDANCE WITH LADOTD 2006 STANDARD SPECIFICATION 728. CASING PIPE DIAMETER SHALL BE NO GREATER THAN 48".

2. THE CASING PIPE SHALL BE MADE OF SMOOTH WALL WELDED CARBON STEEL PIPE CONFORMING TO ASTM A139, GRADE B HAVING A MINIMUM YIELD STRENGTH OF 38,000 PSI. CASING PIPE SHALL BE COATED WITH COAL TAR EPOXY-POLYAMIDE PAINT, IN ACCORDANCE WITH THE LADOTD 2006 STANDARD SPECIFICATION 1008.04.

3. TRENCH EXCAVATION TO FACILITATE THE JACKING CONSTRUCTION PER LADOTD 2006 STANDARD SPECIFICATION 728 (EMBANKMENT MAY BE STEPPED FOR SAFETY).

COASTAL PROTECTION AND
RESTORATION AUTHORITY
450 LAUREL STREET
BATON ROUGE, LOUISIANA 70801

BY					
DESCRIPTION					
REV.	DATE				

DRAWN BY: KRISTI CANTU   DESIGNED BY: RUDOLPH SIMONEAUX, P.E.

LAKE HERMITAGE
MARSH CREATION

STATE PROJECT NUMBER: BA-42

FEDERAL PROJECT NUMBER:

APPROVED BY: JERRY CARROLL, P.E.

MISSISSIPPI RIVER
LEVEE AND HIGHWAY
CROSSING SECTIONS

DATE: SEPTEMBER 2011

SHEET 13 OF 23

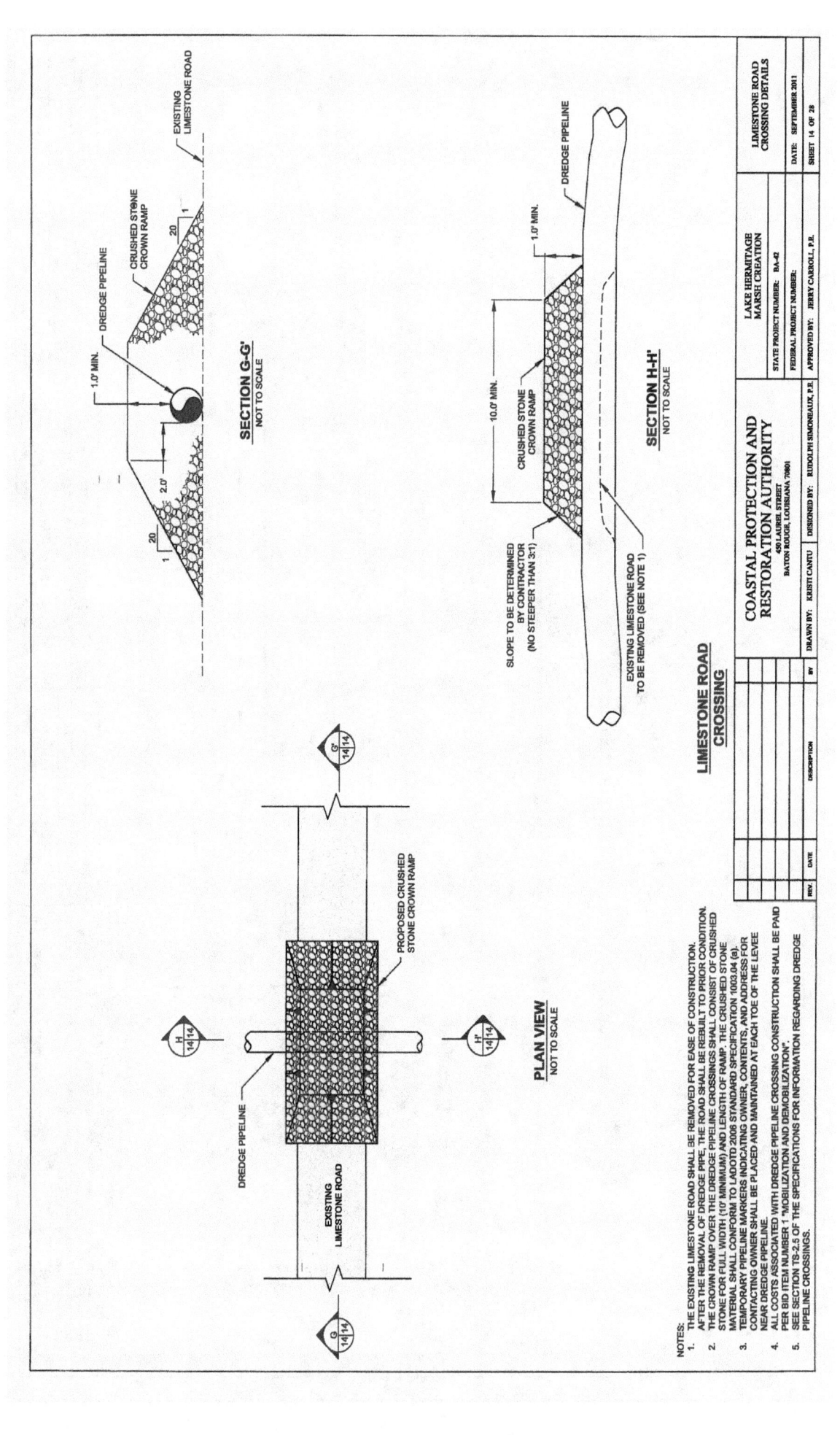

SECTION G-G'
NOT TO SCALE

1.0' MIN.

2.0'

20
1

DREDGE PIPELINE

CRUSHED STONE
CROWN RAMP

EXISTING
LIMESTONE ROAD

SECTION H-H'
NOT TO SCALE

1.0' MIN.

10.0' MIN.

DREDGE PIPELINE

CRUSHED STONE
CROWN RAMP

SLOPE TO BE DETERMINED
BY CONTRACTOR
(NO STEEPER THAN 3:1)

EXISTING LIMESTONE ROAD
TO BE REMOVED (SEE NOTE 1)

PLAN VIEW
NOT TO SCALE

DREDGE PIPELINE

PROPOSED CRUSHED
STONE CROWN RAMP

EXISTING
LIMESTONE ROAD

G
14 14

G'
14 14

H
14 14

H'
14 14

NOTES:

1. THE EXISTING LIMESTONE ROAD SHALL BE REMOVED FOR EASE OF CONSTRUCTION. AFTER THE REMOVAL OF DREDGE PIPE, THE ROAD SHALL BE REBUILT TO PRIOR CONDITION.

2. THE CROWN RAMP OVER THE DREDGE PIPELINE CROSSINGS SHALL CONSIST OF CRUSHED STONE FOR FULL WIDTH (10' MINIMUM) AND LENGTH OF RAMP. THE CRUSHED STONE MATERIAL SHALL CONFORM TO LADOTD 2006 STANDARD SPECIFICATION 1003.04 (a).

3. TEMPORARY PIPELINE MARKERS INDICATING OWNER, CONTENTS, AND ADDRESS FOR CONTACTING OWNER SHALL BE PLACED AND MAINTAINED AT EACH TOE OF THE LEVEE NEAR DREDGE PIPELINE.

4. ALL COSTS ASSOCIATED WITH DREDGE PIPELINE CROSSING CONSTRUCTION SHALL BE PAID PER BID ITEM NUMBER 1 "MOBILIZATION AND DEMOBILIZATION".

5. SEE SECTION TS-2.5 OF THE SPECIFICATIONS FOR INFORMATION REGARDING DREDGE PIPELINE CROSSINGS.

LIMESTONE ROAD
CROSSING

COASTAL PROTECTION AND
RESTORATION AUTHORITY
450 LAUREL STREET
BATON ROUGE, LOUISIANA 70801

LAKE HERMITAGE
MARSH CREATION

STATE PROJECT NUMBER: BA-42

FEDERAL PROJECT NUMBER:

DRAWN BY: KRISTI CANTU

DESIGNED BY: RUDOLPH SIMONEAUX, P.E.

APPROVED BY: JERRY CARROLL, P.E.

LIMESTONE ROAD
CROSSING DETAILS

DATE: SEPTEMBER 2011

SHEET   14   OF   28

REV.	DATE	DESCRIPTION	BY

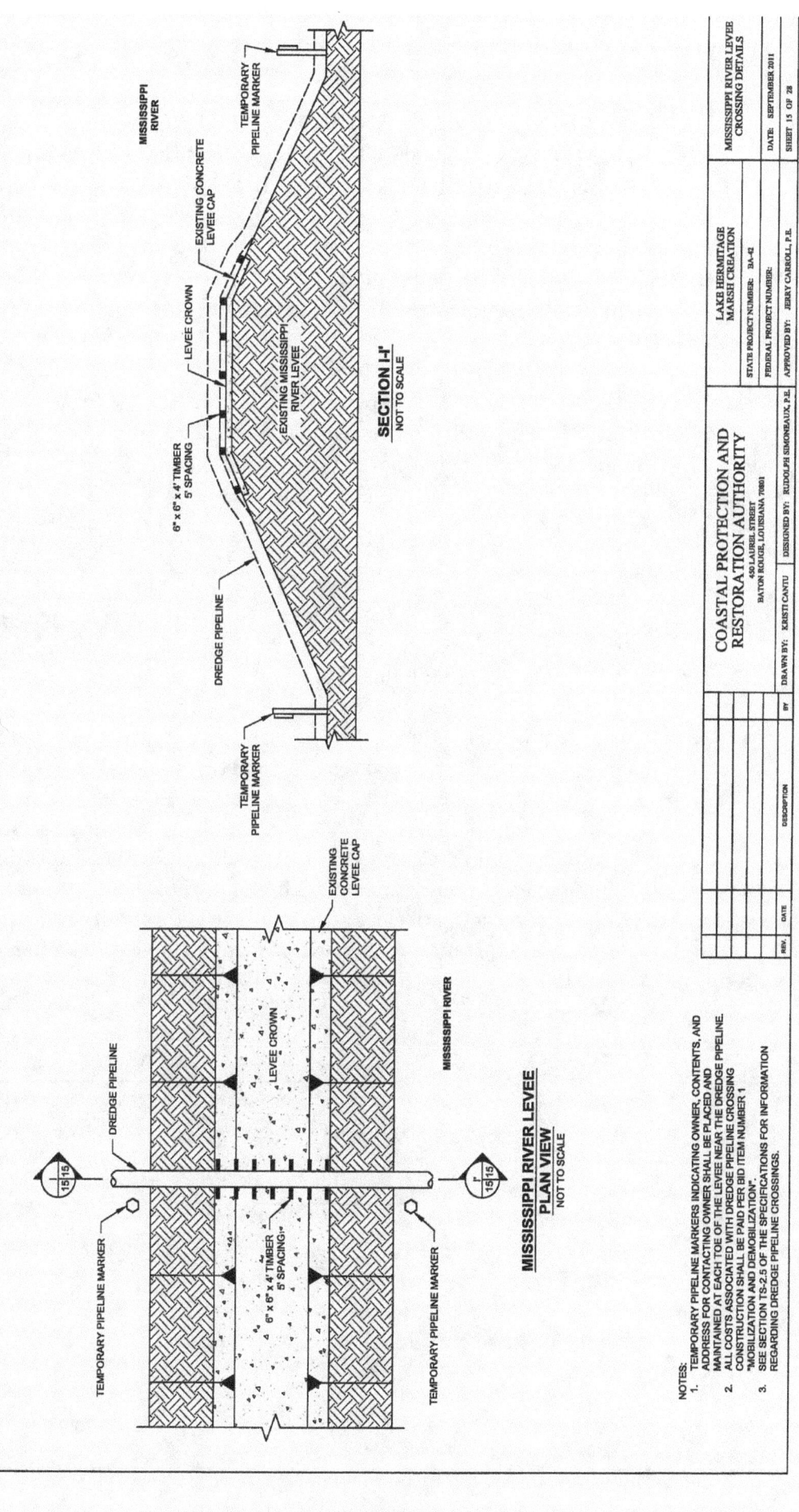

SECTION I-I"
NOT TO SCALE

MISSISSIPPI RIVER

TEMPORARY PIPELINE MARKER

EXISTING CONCRETE LEVEE CAP

LEVEE CROWN

6" x 6" x 4' TIMBER 5' SPACING

DREDGE PIPELINE

EXISTING MISSISSIPPI RIVER LEVEE

TEMPORARY PIPELINE MARKER

MISSISSIPPI RIVER LEVEE
PLAN VIEW
NOT TO SCALE

EXISTING CONCRETE LEVEE CAP

MISSISSIPPI RIVER

DREDGE PIPELINE

LEVEE CROWN

6" x 6" x 4' TIMBER 5' SPACING

TEMPORARY PIPELINE MARKER

TEMPORARY PIPELINE MARKER

NOTES:
1. TEMPORARY PIPELINE MARKERS INDICATING OWNER, CONTENTS, AND ADDRESS FOR CONTACTING OWNER SHALL BE PLACED AND MAINTAINED AT EACH TOE OF THE LEVEE NEAR THE DREDGE PIPELINE.
2. ALL COSTS ASSOCIATED WITH DREDGE PIPELINE CROSSING CONSTRUCTION SHALL BE PAID PER BID ITEM NUMBER 1 "MOBILIZATION AND DEMOBILIZATION".
3. SEE SECTION TS-2.5 OF THE SPECIFICATIONS FOR INFORMATION REGARDING DREDGE PIPELINE CROSSINGS.

COASTAL PROTECTION AND
RESTORATION AUTHORITY
450 LAUREL STREET
BATON ROUGE, LOUISIANA 70801

LAKE HERMITAGE
MARSH CREATION

STATE PROJECT NUMBER: BA-42

FEDERAL PROJECT NUMBER:

MISSISSIPPI RIVER LEVEE
CROSSING DETAILS

DATE: SEPTEMBER 2011

SHEET 15 OF 28

| DRAWN BY: KRISTI CANTU | DESIGNED BY: RUDOLPH SIMONEAUX, P.E. | APPROVED BY: JERRY CARROLL, P.E. |

| REV. | DATE | DESCRIPTION | BY |

FINDING OF NO SIGNIFICANT IMPACT
LAKE HERMITAGE MARSH CREATION PROJECT (BA-42)
PLAQUEMINES PARISH, LOUISIANA

The U.S. Fish and Wildlife Service (Service) is proposing to construct the Lake Hermitage Marsh Creation Project (BA-42), located in Plaquemines Parish, Louisiana. The project is funded through the Coastal Wetlands Planning, Protection and Restoration Act and was authorized for construction on January 21, 2009.

An Environmental Assessment (EA) has been prepared which addresses the Preferred Alternative and a No Action alternative. The purpose of the proposed project is to create 601 acres of emergent marsh by hydraulically dredging bottom sediments and placing that material in shallow open water and fragmented marsh areas. The project will restore approximately 7,400 feet of the eastern shoreline of Lake Hermitage. In addition, approximately 7,300 feet of terraces will be constructed.

Copies of the draft EA were distributed to all pertinent local, state and Federal agencies, and public coastal Louisiana restoration groups in November, 2008. After a 30-day comment period, pertinent comments were incorporated into the final EA.

The Preferred Alternative of creating marsh in shallow open water areas was selected because it will restore emergent marsh in the project area and result in a net gain of 447 acres of marsh compared to the No Action Alternative.

Based on my review and evaluation of the enclosed EA, I have determined that the Lake Hermitage Marsh Creation Project is not a major Federal action which would significantly affect the quality of the human environment within the meaning of Section 102(2)(C) of the National Environmental Policy Act of 1969. Accordingly, preparation of an Environmental Impact Statement on the proposed action is not required.

_____
Acting Field Supervisor
Louisiana Ecological Services Office

November 8, 2011
_____
Date

Reference:

Final Environmental Assessment, dated November 2011

Enclosure
03129196 FWM 246   ENVIRONMENTAL QUALITY

United States Department of the Interior

FISH AND WILDLIFE SERVICE
646 Cajundome Blvd.
Suite 400
Lafayette, Louisiana 70506

April 16, 2012

Memorandum for the Record

From: Darryl Clark, Louisiana Ecological Services, FWS, Lafayette, LA

Subject: Lake Hermitage Marsh Creation Project (BA-42) Finding of No Significant Impact
            (FONSI).

On November 8, 2011, the Fish and Wildlife Service issued a Finding of No Significant Impact
(FONSI) for the Preferred Alternative for the Lake Hermitage Marsh Creation Project (BA-42) in
Plaquemines Parish, Louisiana. The FONSI includes the "Alternative to Terraces" set forth in the
Preferred Alternative in the corresponding Environmental Assessment.

FINDING OF NO SIGNIFICANT IMPACT
LOUISIANA OYSTER CULTCH PROJECT
COASTAL LOUISIANA

The State of Louisiana's Coastal Protection and Restoration Authority, Oil Spill Coordinator's Office, Department of Environmental Quality, Department of Wildlife and Fisheries and Department of Natural Resources, along with the other state and federal natural resource trustees, including the Department of the Interior (collectively, the Trustees), are proposing to implement the Louisiana Oyster Cultch Project. The project involves (1) the placement of oyster cultch onto approximately 850 acres of public oyster seed grounds throughout coastal Louisiana and (2) construction of an oyster hatchery facility that would serve to improve existing oyster hatchery operations and produce supplemental larvae and seed. The proposed cultch placement locations include public oyster seed grounds at 3-Mile Bay, Drum Bay, Lake Fortuna, South Black Bay, Hackberry Bay and Sister Lake. The hatchery facility would be located at the Sea Grant oyster hatchery at the Louisiana Department of Wildlife and Fisheries facility on Grand Isle, Louisiana. The project is an early restoration project funded as part of the *Deepwater Horizon* Natural Resource Damage Assessment and Restoration process in accordance with the "Framework for Early Restoration Addressing Injuries Resulting from the Deepwater Horizon Oil Spill." It was one of several projects proposed for implementation by the Trustees in a Draft Phase I Early Restoration Plan and Environmental Assessment to accelerate restoration, and represents an initial step toward the restoration of natural resources injured by the *Deepwater Horizon* oil spill. The Trustees have considered public comments on that plan and now intend to finalize the selection of the Louisiana Oyster Cultch Project as an early restoration action.

Under the Oil Pollution Act of 1990, damages recovered from parties responsible for natural resource injuries are used to restore, replace, rehabilitate and/or acquire the equivalent of the injured natural resources. *See* 33 U.S.C. 2706. When federal trustees are involved, these restoration activities are subject to the requirements of the National Environmental Policy Act (NEPA), 42 U.S.C 4321 *et seq.*.

An Environmental Assessment (EA) has been prepared which addresses the Proposed Action and a No Action alternative. The purpose of the Proposed Action is to begin to restore, replace, rehabilitate, and/or acquire the equivalent of Louisiana's oyster resources. We have prepared the Final EA and Finding of No Significant Impact after considering input from the public during the comment period for the Draft Phase I Early Restoration Plan/EA.

The Proposed Action was selected because it will result in more efficient recovery or restoration of oyster resources in coastal Louisiana compared to the No Action Alternative.

DETERMINATION

Based on review and evaluation of the EA, it has been determined that the Louisiana Oyster Cultch Project is not a major federal action which would significantly affect the quality of the human environment. Accordingly, the preparation of an Environmental Impact Statement is not required. *See* 42 U.S.C. 4332(2)(C).

Reasons:
- The proposed action may result in minimal and short-term impacts to water quality and habitat during construction of the hatchery and placement of the oyster cultch. However, any potential water quality impacts would be minor and localized to the period of construction/implementation. The hatchery includes a water filtration system. The only addition to the water in the hatchery system is algae, which is taken up by the oyster larvae and broodstock, resulting in no adverse impacts to water quality. Because oysters are filter feeders, the hatchery operation would likely improve water as water passes through the system.

- All necessary permits have been, or will be, obtained and permit conditions, regulations, policies and laws will be followed. The project is expected to receive permitting approval under the New Orleans District Corps of Engineers Programmatic General Permit (PGP) for the Louisiana Coastal Zone. Recent oyster cultch placement projects in Louisiana have been permitted under the PGP.
- The proposed project would place oyster cultch material onto existing public oyster seed grounds.
- A complete review of this project under Section 106 of the National Historic Preservation Act will be completed prior to project implementation.
- No adverse effects to marine mammals are expected from this project.
- Endangered Species Act Section 7 consultation has been completed and no adverse effects to endangered species are expected.
- Essential Fish Habitat (EFH) consultation has been completed and it was determined that this project would not adversely affect EFH, and overall would likely benefit federally managed fisheries species.
- The Louisiana Department of Natural Resources (LDNR) has evaluated the project and determined the project to be broadly consistent with the Louisiana Coastal Resource Program (LCRP). LDNR will issue a final determination upon receipt of the final consistency determination or Coastal Use Permit application for the project.
- No significant adverse cumulative impacts are anticipated from implementation of this project. The project is consistent with ongoing actions of the State of Louisiana.
- Copies of the draft EA for this project were made available to the public through a Federal Register notice on December 14, 2011. *See* Deepwater Horizon Oil Spill; Draft Phase I Early Restoration Plan and Environmental Assessment, 76 Fed. Reg. 78,016 – 78, 018 (Dec. 14, 2011). Public comments on the draft EA were taken during a 60 day public comment period extending from December 14, 2011 to February 14, 2012. Public comments that were received during this period have been considered and incorporated into the final EA. The Phase I Early Restoration Plan/Environmental Assessment is hereby incorporated by reference.

Date: 4|13|2012

Signature:

Cynthia Dohner
Authorized Official, U.S. Department of the Interior

FINDING OF NO SIGNIFICANT IMPACT
OYSTER CULTCH RESTORATION PROJECT
HANCOCK AND HARRISON COUNTIES, MISSISSIPPI

The Mississippi Department of Environmental Quality, along with the other state and federal natural resource trustees, including the Department of the Interior (collectively, the Trustees), is proposing to construct the Mississippi Oyster Cultch Early Restoration Project located in the Mississippi Sound, Hancock and Harrison Counties, Mississippi. The project involves restoring 1,430 acres of oyster cultch areas in the marine waters of the State of Mississippi. The project is an early restoration project funded as part of the *Deepwater Horizon* Natural Resource Damage Assessment and Restoration process in accordance with the "Framework for Early Restoration Addressing Injuries Resulting from the Deepwater Horizon Oil Spill." It was one of several projects proposed for implementation by the Trustees in a Draft Phase I Early Restoration Plan and Environmental Assessment to accelerate restoration, and represents an initial step toward the restoration of natural resources injured by the *Deepwater Horizon* oil spill. The Trustees have considered public comments on that plan and now intend to finalize the selection of the Mississippi Oyster Cultch Restoration Project as an early restoration action.

Under the Oil Pollution Act of 1990, damages recovered from parties responsible for natural resource injuries are used to restore, replace, rehabilitate and/or acquire the equivalent of the injured natural resources. *See* 33 U.S.C. 2706. When federal trustees are involved, these restoration activities are subject to the requirements of the National Environmental Policy Act (NEPA) 42 U.S.C. 4321 *et seq.*.

An Environmental Assessment (EA) has been prepared which addresses the Proposed Action and a No Action alternative. The purpose of the Proposed Action is to begin to restore, replace, rehabilitate and/or acquire the equivalent of Mississippi's oyster resources. We have prepared the Final EA and Finding of No Significant Impact after considering input from the public during the comment period for the Draft Phase I Early Restoration Plan/EA.

The Proposed Action was selected because it will result in more efficient recovery or restoration of oyster secondary production in the Mississippi Sound compared to the No Action Alternative.

DETERMINATION
Based on review and evaluation of the EA, it has been determined that the Mississippi Oyster Cultch Restoration Project is not a major federal action which would significantly affect the quality of the human environment. Accordingly, the preparation of an Environmental Impact Statement is not required. *See* 42 U.S.C. 4332(2)(C).

Reasons:
- Implementation of the proposed action may result in minimal and short-term impacts to water quality and habitat due to placement of the oyster cultch needed to restore and enhance oyster habitat.
- All necessary permits have been obtained and permit conditions, regulations, policies and laws will be followed.
- The proposed project would only place cultch material on existing reef footprints and would not replace soft bottomed habitats.
- The project received permitting approval from the U.S. Army Corp of Engineers through the Nationwide Permit program indicating the impacts associated with the project are minor in nature, provided that permit conditions are adhered to.
- A complete review of this project under Section 106 of the National Historic Preservation Act will be completed prior to project implementation.

- No adverse effects to marine mammals are expected from the project.
- Endangered Species Act Section 7 consultation has been completed and no adverse effects to endangered species are expected.
- Essential Fish Habitat consultation has been completed and no adverse effects are anticipated.
- The Mississippi Department of Marine Resources has confirmed that the project is consistent with Mississippi's Coastal Zone Management Plan.
- No significant adverse cumulative impacts are anticipated from implementation of this project.
- Copies of the draft EA for this project were made available to the public through a Federal Register notice on December 14, 2011. *See* Deepwater Horizon Oil Spill; Draft Phase I Early Restoration Plan and Environmental Assessment, 76 Fed. Reg. 78,016 – 78, 018 (Dec. 14, 2011). Public comments on the draft EA were taken during a 60 day public comment period extending from December 14, 2011 to February 14, 2012. Public comments that were received during this period have been considered and incorporated into the final EA. The Phase I Early Restoration Plan/Environmental Assessment is hereby incorporated by reference.

Date: 4/13/2012

Signature:

Cynthia Dohner
Authorized Official, U.S. Department of the Interior

FINDING OF NO SIGNIFICANT IMPACT
ARTIFICIAL REEF HABITAT RESTORATION PROJECT
MISSISSIPPI

The Mississippi Department of Environmental Quality, along with the other state and federal natural resource trustees, including the Department of the Interior (collectively, the Trustees), is proposing to implement the Mississippi Artificial Reef Habitat Restoration Project located on 67 existing nearshore artificial reefs in Mississippi waters. The project consists of the restoration and enhancement of these existing reefs that are approximately 3 acres in size (201 acres in total) using crushed limestone. The project is an early restoration project funded as part of the *Deepwater Horizon* Natural Resource Damage Assessment and Restoration process in accordance with the "Framework for Early Restoration Addressing Injuries Resulting from the Deepwater Horizon Oil Spill." It was one of several projects proposed for implementation by the Trustees in a Draft Phase I Early Restoration Plan and Environmental Assessment to accelerate restoration, and represents an initial step toward the restoration of natural resources injured by the *Deepwater Horizon* oil spill. The Trustees have considered public comments on that plan and now intend to finalize the selection of the Mississippi Artificial Reef Habitat Restoration Project as an early restoration action.

Under the Oil Pollution Act of 1990, damages recovered from parties responsible for natural resource injuries are used to restore, replace, rehabilitate and/or acquire the equivalent of the injured natural resources. *See* 33 U.S.C. 2706. When federal trustees are involved, these restoration activities are subject to the requirements of the National Environmental Policy Act (NEPA), 42 U.S.C. 4321 *et seq.*.

An Environmental Assessment (EA) has been prepared which addresses the Proposed Action and a No Action alternative. The purpose of the Proposed Action is to begin to restore, replace, rehabilitate and/or acquire the equivalent of Mississippi's secondary production of invertebrate and infaunal and epifaunal biomass. We have prepared the Final EA and Finding of No Significant Impact after considering input from the public during the comment period for the Draft Phase I Early Restoration Plan EA.

The Proposed Action was selected because it will result in more efficient recovery or restoration of infaunal and epifaunal biomass secondary productivity compared to the No Action Alternative.

DETERMINATION
Based on review and evaluation of the EA, it has been determined that the Mississippi Artificial Reef Habitat Restoration Project is not a major federal action which would significantly affect the quality of the human environment. Accordingly, the preparation of an Environmental Impact Statement is not required. *See* 42 U.S.C. 4332(2)(C).

Reasons:
- Adverse biological impacts would be temporary. Short-term disturbances to the water column and benthic organisms may occur when the project is implemented. Overall the project would result in an improved marine ecosystem. Nearshore artificial reefs would provide valuable hardbottom habitat to help restore secondary production of infaunal and epifaunal biomass.
- All necessary permits have been obtained and permit conditions, regulations, policies and laws will be followed.
- All efforts would be made to avoid existing environmentally sensitive areas such as oyster reefs, emergent and submerged aquatic vegetation, and other live bottom communities.
- The project received permitting approval from the U.S. Army Corp of Engineers through the Nationwide Permit program indicating the impacts associated with the project are minor in nature, provided that permit conditions are adhered to.

- A complete review of this project under Section 106 of the National Historic Preservation Act will be completed prior to project implementation.
- No adverse effects to marine mammals are expected from the project.
- Endangered Species Act Section 7 consultation has been completed and no adverse effects to endangered species are expected.
- Essential Fish Habitat (EFH) consultation has been completed and it was determined that this project would not adversely affect EFH, and overall would likely benefit federally managed fisheries species.
- The Mississippi Department of Marine Resources has confirmed that the project is consistent with Mississippi's Coastal Zone Management Plan.
- No significant adverse cumulative impacts are anticipated from implementation of this project.
- Copies of the draft EA for this project were made available to the public through a Federal Register notice on December 14, 2011. *See* Deepwater Horizon Oil Spill; Draft Phase I Early Restoration Plan and Environmental Assessment, 76 Fed. Reg. 78,016 – 78, 018 (Dec. 14, 2011). Public comments on the draft EA were taken during a 60 day public comment period extending from December 14, 2011 to February 14, 2012. Public comments that were received during this period have been considered and incorporated into the final EA. The Phase I Early Restoration Plan/Environmental Assessment is hereby incorporated by reference.

Date: 4/13/2012

Signature: Cynthia Dohner
Authorized Official, U.S. Department of the Interior

FINDING OF NO SIGNIFICANT IMPACT
BOAT RAMP ENHANCEMENT AND CONSTRUCTION PROJECT
ESCAMBIA COUNTY, FLORIDA

The Florida Department of Environmental Protection and the Florida Fish and Wildlife Conservation Commission, along with the other state and federal natural resource trustees, including the Department of the Interior (collectively, the Trustees), are proposing to implement the Florida Boat Ramp Enhancement and Construction Project at four different sites, all in Escambia County. This project is consistent with the goal of restoring or replacing human use service losses resulting from the *Deepwater Horizon* oil spill. This project would entail repairing an existing boat ramp in Pensacola Bay (Navy Point Park Public Boat Ramp) and construction of a new boat ramp facility in Pensacola Bay (Mahogany Mill Public Boat Ramp). The project also includes repairing and modifying an existing boat ramp in Perdido Bay (Galvez Landing Public Boat Ramp) and construction of a new boat ramp facility in Perdido River (Perdido Public Boat Ramp). Visitor information kiosks would be installed to provide environmental education to boaters regarding water quality and sustainable practices for utilization of marine/estuarine/coastal resources in Florida. The project is an early restoration project funded as part of the *Deepwater Horizon* Natural Resource Damage Assessment and Restoration process in accordance with the "Framework for Early Restoration Addressing Injuries Resulting from the Deepwater Horizon Oil Spill." It was one of several projects proposed for implementation by the Trustees in a Draft Phase I Early Restoration Plan and Environmental Assessment to accelerate restoration, and represents an initial step toward addressing the reduced quality and quantity of recreational activities (e.g., boating and fishing) tied to natural resources injured by the *Deepwater Horizon* oil spill. The Trustees have considered public comments on that plan and now intend to finalize the selection of the Florida Boat Ramp Enhancement and Construction Restoration Project as an early restoration action.

Under the Oil Pollution Act of 1990, damages recovered from parties responsible for natural resource injuries are used to restore, replace, rehabilitate and/or acquire the equivalent of the injured natural resources. *See* 33 U.S.C. § 2706. When federal trustees are involved, these restoration activities are subject to the requirements of National Environmental Policy Act (NEPA), 42 U.S.C. § 4321. *et seq.*.

An Environmental Assessment (EA) has been prepared which addresses the Proposed Action and a No Action alternative. The purpose of the Proposed Action is to begin to restore, replace, rehabilitate, and/or acquire the equivalent of recreational service losses in Florida attributable to the *Deepwater Horizon* oil spill. We have prepared the Final EA and Finding of No Significant Impact after considering input from the public during the comment period for the Draft Phase I Early Restoration Plan/EA.

The Proposed Action was selected because it will provide local boaters with access to public waterways and water recreational activities (including fishing, diving, water-skiing, SCUBA diving, and cruising), resulting in more efficient recovery or restoration of human use of natural resources compared to the No Action Alternative.

DETERMINATION
Based on review and evaluation of the EA, it has been determined that the Florida Boat Ramp Enhancement and Construction Restoration Project is not a major federal action which would significantly affect the quality of the human environment. Accordingly, preparation of an Environmental Impact Statement on the proposed action is not required. *See* 42 U.S.C. § 4332(2)(C).

Reasons:
- The proposed boat ramp locations, whether new construction or existing ramps, are all located in developed areas. Boat ramp construction and operation would cause only minimal alteration and/or damage to habitats.

- Escambia County is not listed as one of the 36 Florida coastal and inland counties in which manatees regularly occur. Manatees would not be attracted to the area of the boat ramps due to the lack of submerged vegetation for foraging at the sites. The project sites are not adjacent to manatee protection zones so the risk of collision around the boat ramps is low.
- Increases in boating opportunities and recreational fishing are not expected to adversely impact fish populations. The number of new trips generated by the construction and modification of these four boat ramps will not be significant in the context of the total number of trips generated by all access points in Florida.
- The boat ramps would be constructed on already developed sites where it is not likely that nesting shore- and seabirds would be impacted.
- All necessary permits have been, or will be obtained, and permit conditions, regulations, policies and laws will be followed.
- Impacts to water quality are expected to be minimal. All permit conditions requiring mitigation measures for siltation, erosion, turbidity and release of chemicals will be strictly adhered to.
- A complete review of this project under Section 106 of the National Historic Preservation Act will be completed prior to project implementation.
- Endangered Species Act Section 7 consultation has been completed and no adverse effects to endangered species are expected.
- Essential Fish Habitat (EFH) consultation has been completed and it was determined that this project would not adversely affect EFH.
- The Florida Department of Environmental Protection has confirmed that the Phase I Early Restoration Plan is consistent with the Florida Coastal Zone Management Plan.
- The project would improve access to public waterways, benefitting recreational opportunities.
- No significant adverse cumulative impacts are anticipated from implementation of this project.
- Copies of the draft EA for this project were made available to the public through a Federal Register notice on December 14, 2011. *See* Deepwater Horizon Oil Spill; Draft Phase I Early Restoration Plan and Environmental Assessment, 76 Fed. Reg. 78,016 – 78, 018 (Dec. 14, 2011). Public comments on the draft EA were taken during a 60 day public comment period extending from December 14, 2011 to February 14, 2012. Public comments that were received during this period have been considered and incorporated into the final EA. The Phase I Early Restoration Plan/Environmental Assessment is hereby incorporated by reference.

Date: 4/13/2012

Signature:

Cynthia Dohner
Authorized Official, U.S. Department of the Interior

UNITED STATES FISH AND WILDLIFE SERVICE
ENVIRONMENTAL ACTION STATEMENT

Within the spirit and intent of the Council on Environmental Quality's regulations for implementing the National Environmental Policy Act (NEPA), and other statutes, orders, and policies that protect fish and wildlife resources, I have established the following administrative record and determined that the action of restoring vegetated dune habitat through dune plantings and sand fencing for the Alabama Dune Restoration Cooperative Early Restoration Project meets two resource management categorical exclusions: (3) The construction of new, or the addition of, small structures or improvements, including structures and improvements for restoration of wetland, riparian, instream, or native habitats, which result in no or only minor changes in the use of the affected local area. The following are examples of activities that may be included.
(a) The installation of fences.
(b) The construction of small water control structures.
(c) The planting of seeds or seedlings and other minor revegetation actions.
(d) The construction of small berms or dikes.
(e) The development of limited access for routine maintenance and management purposes.

(11) Natural resource damage assessment restoration plans, prepared under sections 107, 111, and 122(j) of the Comprehensive Environmental Response Compensation and Liability Act (CERCLA); section 311(f)(4) of the Clean Water Act; and the Oil Pollution Act; when only minor or negligible change in the use of the affected areas is planned.

Check One:

__X__ is a categorical exclusion as provided by 51 6 DM 8.5, B (3) and B (11) No further NEPA documentation will therefore be made.

_____ is found not to have significant environmental effects as determined by the attached environmental assessment and finding of no significant impact.

_____ is found to have significant effects and, therefore, further consideration of this action will require a notice of intent to be published in the Federal Register announcing the decision to prepare an EIS.

_____ is not approved because of unacceptable environmental damage, or violation of Fish and Wildlife Service mandates, policy, regulations, or procedures.

_____ is an emergency action within the context of 40 CFR 1506.11. Only those actions necessary to control the immediate impacts of the emergency will be taken. Other related actions remain subject to NEPA review.

Other supporting documents (list):
See attached Deepwater Horizon Oil Spill Phase I Early Restoration Plan and Environmental Assessment for the Alabama Dune Restoration Cooperative Early Restoration Project.

Signature Approval:

_____   _____
Regional Director   Date

UNITED STATES FISH AND WILDLIFE SERVICE
ENVIRONMENTAL ACTION STATEMENT

Within the spirit and intent of the Council on Environmental Quality's regulations for implementing the National Environmental Policy Act (NEPA), and other statutes, orders, and policies that protect fish and wildlife resources, I have established the following administrative record and determined that the action of restoring vegetated dune habitat through dune plantings for the Florida (Pensacola Beach) Dune Early Restoration Project meets two resource management categorical exclusions:

(3) The construction of new, or the addition of, small structures or improvements, including structures and improvements for restoration of wetland, riparian, instream, or native habitats, which result in no or only minor changes in the use of the affected local area. The following are examples of activities that may be included.
(a) The installation of fences.
(b) The construction of small water control structures.
(c) The planting of seeds or seedlings and other minor revegetation actions.
(d) The construction of small berms or dikes.
(e) The development of limited access for routine maintenance and management purposes.

(11) Natural resource damage assessment restoration plans, prepared under sections 107, 111, and 122(j) of the Comprehensive Environmental Response Compensation and Liability Act (CERCLA); section 311(f)(4) of the Clean Water Act; and the Oil Pollution Act; when only minor or negligible change in the use of the affected areas is planned.

Check One:

___X___ is a categorical exclusion as provided by 51 6 DM 8.5, B (3) and B (11) No further NEPA documentation will therefore be made.

_____ is found not to have significant environmental effects as determined by the attached environmental assessment and finding of no significant impact.

_____ is found to have significant effects and, therefore, further consideration of this action will require a notice of intent to be published in the Federal Register announcing the decision to prepare an EIS.

_____ is not approved because of unacceptable environmental damage, or violation of Fish and Wildlife Service mandates, policy, regulations, or procedures.

_____ is an emergency action within the context of 40 CFR 1506.11. Only those actions necessary to control the immediate impacts of the emergency will be taken. Other related actions remain subject to NEPA review.

Other supporting documents (list):
See attached Deepwater Horizon Oil Spill Phase I Early Restoration Plan and Environmental Assessment for the Florida (Pensacola Beach) Dune Early Restoration Project.

Signature Approval:

Cynthia K. Dohner 4/10/2012
Director/Regional   Date

UNITED STATES DEPARTMENT OF COMMERCE
National Oceanic and Atmospheric Administration
NATIONAL MARINE FISHERIES SERVICE
Southeast Regional Office
263 13th Avenue South
St. Petersburg, Florida 33701-5511
(727)824-5317; FAX (727) 824-5300
http://sero.nmfs.noaa.gov/

February 14, 2012   F/SER4:DD

MEMORANDUM FOR: Jeff Shenot
                    NOAA Restoration Center

FROM: Virginia M. Fay
                    Assistant Regional Administrator, Habitat Conservation Division

SUBJECT: Essential Fish Habitat (EFH) review of the Phase I Early
                    Restoration Projects

This responds to your February 2, 2012, electronic mail correspondence requesting an EFH
review of the subject action. The natural resource Trustees for the BP/Deepwater Horizon
incident propose to conduct eight early restoration projects in the coastal areas of Louisiana,
Mississippi, Alabama, and Florida.

As specified in the Magnuson-Stevens Fishery Conservation and Management Act (Magunson-
Stevens Act), EFH consultation is required for federal actions which may adversely affect EFH.
Representing the federal Trustees for the subject action, the NOAA Restoration Center
determined the proposed actions would not adversely affect EFH, and overall, would likely
benefit federally managed fishery species. It is important to note the EFH assessment identifies
and analyzes the conversion of habitat types that may result from the proposed restoration
activities. The assessment also identifies best management practices and measures to minimize
indirect impacts to EFH. Based on our review of the preliminary projects, we do not have any
EFH Conservation Recommendations to provide pursuant to Section 305(b)(2) of the Magnuson-
Stevens Act at this time. Please be advised that further consultation on this matter is not
necessary unless future modifications are proposed.

Future EFH Assessments should reflect recent actions by the Gulf of Mexico Fishery
Management Council, and approved by the NMFS, removing certain species from federal
management under the Magnuson-Stevens Act. The fishery management plan for stone crab in
the Gulf of Mexico was repealed effective October 24, 2011. Additionally, the following were
removed from fishery management units and plans effective in January 2012: dog snapper,
mahogany snapper, schoolmaster, misty grouper, red hind, rock hind, blackline tilefish, anchor
tilefish, dwarf sand perch, sand perch, bluefish, cero, dolphin, little tuny, and slipper lobster.

Questions regarding EFH in the Southeast Region may be directed to David Dale at 727-824-
5317 or by email at david.dale@noaa.gov.

www.ingramcontent.com/pod-product-compliance
Lightning Source LLC
Chambersburg PA
CBHW080803180526
45168CB00006B/2315